RESURRECTION
OF
THE BLUE PLANET

RESURRECTION
OF
THE BLUE PLANET

The Plan to Reduce Global Warming

Richard Navarro, Ph.D.
An inhabitant of the Planet Earth

To order additional copies of this book, contact:
Xlibris Corporation
1-888-795-4274
www.Xlibris.com
Orders@Xlibris.com
57917

Introduction

**Figure 1. President Theodore Roosevelt.
(Compliments of Wikipedia)**

In 1908 President Theodore Roosevelt said . . .

> "We have become great because of the lavish use of our resources and we have just reason to be proud of our growth... But the time has come to inquire seriously what will happen when our forests are gone, when the coal, iron, the oil and the gas are exhausted." (Smithsonian, Jan. 2008, p. 50).

Over 100 years ago, this great statesman identified a potential problem as to the depletion of our resources. At the time, no one knew we would eventually have to confront the larger issue of dealing with the effects of our high consumption of natural resources. We have been lavish and as a country we have thrived but we are paying a price that will only become greater the longer we delay in paying off this debt. The damage done to the planet is similar to running up our credit cards to the max until we can only afford to pay the minimums. Every month we get further and further behind as the high interest adds to the balance in spite of our efforts to pay down the debt of global warming. It is time that we all take a serious look at this problem. We need to set up a budget that will allow us to stop the growth of the problem and begin to reduce the overwhelming debt we each owe to our Creator, to each other, to the other inhabitants of Earth, and to this marvelous planet that has graciously allowed us to thrive. This book provides a sample of simple ideas that individuals of all ages can do to save the planet. No heroic actions are necessary as few of us have the wherewithal or resources to do that. While the big polluters get media attention, little attention is paid to the day to day activities which contribute to global warming. It may take years and billions of dollars to implement large scale projects to reduce greenhouse gases while small changes in daily living by enough people can have a large impact on the environment. Many of these ideas don't cost anything except a little extra effort and can be implemented immediately. If every American reduced their carbon emission by one pound a day, there would be a reduction of over 300 million pounds of carbon emissions every day! Multiply that by 365 days a year and you're talking about a real sustainable reduction in carbon emissions! If you really think about it, except for major tragic events like Chernobyl, Hiroshima, wildfires, volcanic eruptions, and refinery explosions, most of the damage being wrought upon the Earth has been brought about by small things that we each do everyday often without thinking about them. A single cigarette butt thrown out the window is not much in the overall scheme of trashing the environment but check out the hundreds of cigarette butts that accumulate near major intersections in big cities or think how that cigarette butt can start a devastating forest fire! Some people believe that humans really aren't the cause of changes in the global environment as the weather changes we are experiencing is merely another cycle of global change that has occurred throughout the millenniums. According to this train of thought, humans are too small and insignificant to effect global change.

Yet, we know a termite is a tiny creature which can destroy an entire home in a relatively short period of time. Any single tiny termite can only eat a tiny portion of the house in a given day but when you have thousands of termites eating round the clock, the house is doomed. Similarly, billions of humans depleting resources, trashing our environment, and adding harmful toxins to the environment can easily doom our planet.

Figure 2. Insidious termite damage.

How difficult is it to reduce your carbon emissions by one pound or more a day? Here are some examples:

1) Reduce your driving by 40 miles a month (or about 1.33 miles per day) = 1.87 pounds less per day depending on your vehicle.
2) Reduce your use of heating oil by 100 gallons a year (or about 2/3 of a gallon a day over a 5 month heating season) = 5.6 pounds less a day
3) Change 4 75 watt standard light bulbs to energy efficient fluorescent bulbs = 1 pound less per day
4) Recycle all your magazines = 1 pound less per day
5) Plant 2 trees = 1.6 pounds less per day

6) Switch from toxic laden common household cleaners to Shaklee Get Clean household and laundry products = 1.35 pounds less per day

7) Reduce your consumption of fast food hamburgers to less than 2 times a week = 1 pound less per day

8) Connect your household electrical devices to a power strip and turn them off when you're not using them. = 2 pounds less per day.

9) Use cold water to wash your clothes instead of hot water = 1.35 pounds less per day.

10) Reduce the temperature on your hot water heater to 120 degrees F = 1.50 pounds less per day.

These are easy, aren't they? This book will give you lots of other ideas which you can use to fit your own particular situation and to pay down your carbon debt.

The single most important thing that each of us can do to reverse global warming and preserve our environment for future generations is to become an active participant in taking care of Mother Earth. This means each of us has to **demonstrate that we care** about the impact of our daily habits so we can make a positive contribution to the environment. Just tightening our gas cap when we fill up can reduce the estimated *147 MILLION GALLONS of wasted gas which is evaporated with defective or absent gas caps.* The simple act of tightening your gas cap may save you money, reduce the amount of gas purchased, and reduce the amount of gas entering the atmosphere. How hard is that?

The first step in changing is to admit that there is a problem and that each of us is part of that problem. At one time or another each of us is guilty of discarding something that could be recycled. Each of us is guilty of turning up (or down) the thermostat. Each of us is guilty of driving too fast, or not planning our errands for maximum fuel and time efficiency. No one is perfect, after all we're humans! *We can each do better if we want to change.* We know that change is an evolutionary process as few people can radically change overnight the habits they've been living with for years. This book will help you make the changes you are comfortable making right now to help reduce your carbon emissions.

How do we really know that there is a problem anyway? All of us have read or heard so many conflicting media stories about different things over the years, we may have developed a hearty sense of skepticism. Some folks don't believe there is global warming simply because former Vice President Al Gore helped bring it to public attention. The fact that he is a democrat somehow casts suspicion on the facts he presents. Others feel that humans are so small and insignificant in the overall scheme of life on this planet that we couldn't possibly destroy it. After all, the Earth is a big planet! Yet, others think global warming is just a natural cycle that the Earth goes through every few millennia. Yet, the average adult has seen more change in their lives in the last 10 years than their grandparents did in their entire lives. For example, 1) it's possible for the average person to travel to the other side of the Earth in less than 24 hours, 2) almost everyone can make an instantaneous phone call anywhere in the world, 3) Live video feeds can be broadcast around the world to show sporting events or disasters.

You don't need anyone else to tell you about global warming. Just think back over your life and ask yourself some simple questions such as . . . Are the days hotter than you remember as a kid? Is there less snow in the winter or more rain? Is there more traffic? Do you feel more stressed everyday? Who do you know that has to stay inside or restrict their outside activities during days of high pollution? Does your car have a film of dust or soot on it the day after you've washed it? Do you still see as many birds or butterflies you did as a kid? What is causing the mysterious disappearance of America's bee population? Do the tomatoes or the peaches taste as good as they used to? Where can you go for a just a few minutes of peace and quiet? If you give some serious thought to the environmental changes around you, it is easy to see that massive changes have occurred.

Figure 3 shows a sample air quality report for Houston, Texas. These are available on line for most large cities. In cities which have a history of air pollution such as Los Angeles, Houston, and Denver some TV stations or newspapers will summarize this information for you. The report will vary at least daily as the air quality is not constant due to changes in the wind, sun, local pollutants, rain, and other factors. For those people who have respiratory problems now, these reports are very important. For those people who don't yet have respiratory problems, they represent the toxins we breathe in everyday. We certainly don't need or want any more toxins to pollute our bodies or impair the health of our children.

Sample Texas Air Quality Summary Table						
Data courtesy of Texas Commission on Environmental Quality (TCEQ)						
G 0-50	**M** **51-100**	**USG** **101-150**	**U** **151-200**	**V** **201-300**	**H** **301-500**	**! Action Day**
Good	Moderate	Unhealthy for Sensitive Groups	Unhealthy	Very Unhealthy	Hazardous	click on icon in city forecast for more details

Houston-Galveston-Brazoria

	Current Air Quality Conditions Monday 25-Aug-08	Today's Forecast Monday 26-Aug-08	Tomorrow's Forecast Tuesday 26-Aug-08
Ozone	AQI observed at: 58 M current conditions yesterday's summary month's archives	**Primary Pollutant !** USG **Health Message:** Active children and adults with lung disease such as asthma should reduce prolonged or heavy heavy exertion outdoors	**Primary Pollutant !** USG **Health Message:** Active children and adults with lung disease such as asthma should reduce prolonged or heavy exertion outdoors.
Particles (PM2.5)	AQI observed at: 63 M current conditions yesterday's summary month's archives	**Secondary Pollutant** M Health Message: Unusually sensitive people should consider reducing prolonged or or heavy exertion	**Secondary Pollutant** M Health Message: Unusually sensitive people should consider reducing prolonged or heavy exertion.
Particles (PM 10)	Data not available	Secondary Pollutant G Health Message: No health impacts are expected in this range	Secondary Pollutant G Health Message: No health impacts are expected in this range.
For additional information visit:			
ozone warnings	Current Ozone Maps		PM2.5 Map
Your Environmental News Flash			

Figure 3. Sample Air Quality Index Report

Then ask yourself one more question.
"Is this the Earth I want my grandchildren to inherit?

Figure 4

The changes you have witnessed were most likely so gradual that you didn't recognize them at the time but when you look back, you can see the changes. There is no denying that the ten hottest years in recorded history have occurred in the last 12 years. There is no question that Hurricane Katrina was one of the most devastating hurricanes to hit the US mainland. There is no question that the typhoons which devastated many areas of southeast Asia in the last few years have brought suffering to millions of people. How do hurricanes develop? Meteorologists know that hurricanes and typhoons are largely the result of the ocean water becoming too warm. At the same time, it is known that changes in the ozone layer and the increase of greenhouse gases have been documented for over 20 years. While there may be some controversy as to other contributing factors, it is known that greenhouse gases prevent the surface air from escaping into outer space; thereby trapping it near the surface where it gets hotter and hotter. Since ¾ of the Earth's surface is water, the ocean surface water temperature increases and the conditions are ripe for a hurricane. Temperature and moisture mapping across the planet documents that the Earth is changing at a very rapid pace. Large chunks of ancient ice in the Arctic and Antarctic are melting away

Figure 5. Hurricane clouds

reducing habitat for polar bears, warming the world climate, and potentially releasing trapped methane gas, a major greenhouse gas. Two thousand eight may be the year with the second smallest mass of arctic ice on record and global weather patterns show a gradual warming trend of colder climates while warm climates are just getting warmer. The effect of global warming and all of its permutations is to create new areas of drought and new areas of flooding. Food production areas are particularly vulnerable as shown by the 2008 flooding in the Midwestern USA.

Figure 6. Drought.

Figure 7 Floods

Another consideration which doesn't get as much attention in the context of global warming includes the significant increases in diseases such as diabetes, high cholesterol, high blood pressure, obesity, asthma, ear infections, hearing loss, allergies, autism, cancer, and mental illness. These medical conditions have all exploded almost exponentially in the last 50 years. In particular the condition known as "metabolic syndrome" which frequently leads to diabetes, heart disease, high blood pressure, hearing loss and obesity have traditionally been viewed as the result of bad eating habits, lack of exercise, and genetics. More current research strongly suggests that environmental pollution is an additional cause that has not been previously considered. It is quite possible that many other conditions such as asthma, allergies, autism, mental illness, and cancer are vested in environmental pollution. These health conditions may be part of the vicious cycle of environmental damage we are all doing to ourselves.

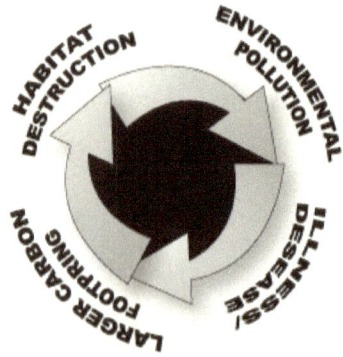

Figure 8. Vicious cycle of how the environment destroys our health.

While the people of industrialized countries have become increasingly obese, many more impoverished countries have rampant malnutrition and starvation. Countries that used to get adequate rainfall to sustain their people and grow food crops are increasingly finding it harder and harder to grow enough food. Poor nutrition, polluted water and air are all hazardous to the health of every person on the planet. Clean drinking water is becoming an increasingly difficult problem for some areas especially for big cities like Los Angeles which gets its water from hundreds of miles away. As water levels decrease and populations increase, geopolitical disputes become more and more likely. While many Americans are stressed due to high gasoline prices, the fact is that a person can live without gasoline. No one can live without water!

One of the major concerns of millions of people is their health. Over 47 million Americans do not have health insurance and millions more spend billions of dollars on medications, diagnostic tests, and doctor visits. The need for better health care has created an ongoing political debate as to the best way to provide health insurance and to stop the upward spiral of health care costs. There is only one way to do this. **Help people stay healthy!** While people will always be getting hurt in accidents, the major causes of the most expensive diseases are preventable through education, better eating habits, more exercise, lifestyle changes, and the reduction of environmental pollution.

Environmental pollution in the home is an issue that everyone should be concerned about because it directly affects the health of you and your family. Studies have shown that the average home is one of the most polluted places. Many products used to build, beautify, and maintain homes are unsafe for people and pets. One study found that housewives that work exclusively at home have a higher probability of cancer than those who work outside the home. The suspected culprit is the high number of toxic chemicals in the home. It has been estimated that the average person is exposed to 1500 chemicals daily. While many of these chemicals are known carcinogens, many more have never been tested. Chemicals are so commonplace that few people take the time to read the labels of their household cleaning products and, if they did, they probably wouldn't understand the implications of what they are reading. Many people just continue to use the same products they have been using without much thought about it. One example is chlorine bleach which is commonly used for many things including household cleaning, laundry, and water purification. The main active ingredient of chlorine bleach is sodium hypochlorite which is mixed with water to various concentration levels ranging from .7 to 5.25%. Sodium hypochlorite is a strong oxidizer which means that it is corrosive and can burn skin and cause eye damage. In addition, lye (which is poison) may be added to bleach as a stabilizer. Two recent studies have concluded that if sodium hypochlorite is mixed with organic chemicals they produce volatile organic compounds which are toxic and probably human carcinogens. Chlorine bleach with lye causes skin irritation or burns due to destruction of the skin oils and tissue and results in a slippery feel on skin. Yet, we continue to wash our children's clothes in bleach which may be absorbed into their skin when they wear them. In 2002, there were about 3300 accidents leading to hospitalizations in British homes due to home toxic poisoning. Within the last few years, more and more people are using chlorine bleach wipes to clean the kitchen counter where they prepare their food so the residue of bleach can easily contaminate our food.

Another example is the use of plastic baby bottles which are often made from bisphenol-a (BPA) which can leach into the milk consumed by the baby. When ingested, BPA may be perceived by the body as the hormone estrogen and create an endocrine disruption to affect normal child development and cause other problems in adults. BPA usage has been linked to increased rates of breast and prostrate cancer, attention deficit hyperactivity disorder, obesity and diabetes. BPA is used extensively to make clear rigid plastic in all types of containers. Unfortunately, the FDA does not accept the numerous recent studies documenting the health risks posed by BPA but rather continues to rely on old out dated studies from the plastics industry. BPA is a known carcinogen. While there are conflicting research studies as to the effects of this chemical on babies, doesn't it make sense to err on the side of caution. I don't want to feed my kids little bits of carcinogens every day, do you?

Once these common household products are used in our homes, they continue to do environmental damage as they pass to landfills, contaminate our soil, or are flushed out to sea where they can contaminate the seafood we eat. It is a tragedy to see the amount of garbage floating in the ocean and to see the effects of this garbage on our marine friends.

While some known carcinogenic herbicides and pesticides like DDT and chlordane can no longer be legally sold in the US, they are still sold and used in other countries which may grow foods consumed in the US. So the carcinogens just come back again and again. These toxins usually take a long time to degrade in the environment and are often washed into the water supply or the ocean. The result is that the toxins become part of the food chain of seafood and, ultimately, to the fish we eat. Consequently, the US Department of Agriculture recommends that no one eat fish more than 1 or 2 times per week as larger fatty fish such as albacore tuna have relatively greater concentrations of mercury. The number and variety of toxic chemicals in our water sources has become so pervasive that millions of frogs and other amphibians are born deformed and some are in jeopardy of extinction. Within the last few years, millions and millions of honey bees in the US have just disappeared. This poses a real problem as honey bees are critical to the pollination of our crops yet no one knows how or why they are disappearing. Could it be that they are the victims of environmental contamination or minute changes in temperature as part of the entire global warming problem?

Millions of people are so concerned about the quality of their drinking water that they spend billions of dollars on bottled water. This results in billions of empty water bottles which don't decompose and fill up our landfills as they are frequently not recycled. This is in spite of statements by public health officials stating that tap water is safe to drink. One reason why so many consumers don't believe public health officials is that experts are human too. As humans, mistakes are made that can effect many people. The recent food recalls of spinach, tomatoes, meat and poultry is testimony to how small mistakes can make hundreds of people sick or dead. As a consumer of water with a scientist's skepticism of research, I wonder about the purity of tap water. Every year one local water authority from which I bought my water mailed statements of water contaminants to their users. These statements listed the EPA guidelines as to the maximum number of parts per million allowed for various contaminants as well as how many were supposedly measured in the water. What I found interesting was that over the course of a few years, the amount of contaminants measured **were always the maximum allowed.** They were never higher and never lower than the maximum by a single part per million. As a scientist, my first reaction was that there was some "fudging" of the numbers as it is almost impossible to get exactly the same measurements every time. In addition, the measurements were probably taken at the main storage facility rather than at my faucet so I had no way of knowing what the actual contaminants were that may have been picked up along the way. So, how do we know that the water is safe? The conclusion of millions of people is that "we don't!" In a market place survey, 20% of shoppers are not confident of the safety of their food and 40% were concerned about the presence of growth hormones in their food.

Although the United States has only 5% of the global population, it has the largest carbon footprint of any nation on the planet as it consumes

20% of the world's energy. Yet, the political leadership of the US has refused to acknowledge and support reductions in global warming until the summer of 2008 when President Bush was reported to say that the flooding in the US Midwest was the result of global warming. Yet, the Kyoto Treaty to reduce global warming has not been ratified (or even considered) by the US Congress. This lack of political leadership appears tied to short term economic policy rather than long term global good. When the devastating rain and flooding destroyed Midwest agricultural production, the US administration finally acknowledged the impact of global warming. Yet, Roberto Dobles, the Energy and Environmental Minister of Costa Rica has stated that "Climate change is the main threat facing humanity . . .". (Yahoo! News, May 12, 2008). While Costa Rica planted 7 million trees (or 1.5 for every citizen) in 2008 in addition to the 5 million trees planted in 2007, the US government does nothing. Environmentally conscious corporations such as the Shaklee Corporation, however, do have their Million Trees Project with the goal of planting a million trees in North America. Shaklee has also supported the planting of a million trees in Kenya as part of Dr. Maathai Mangari's (Nobel Environmental Prize Winner) mission. In addition, developing countries like China and India are increasing their carbon footprint as their economies expand. As the personal wealth of each person grows, they become greater consumers as they can afford more luxuries and each person thereby increases his carbon footprint. Since these two countries have over two billion people, it is inevitable that carbon emissions will continue to rise unless measures are taken to provide for

Fig 9. Carbon footprint.

their needs and wants without increasing carbon emissions. For now, the US has the biggest economy in the world, is producing less and less every year but it's carbon footprint becomes deeper and broader as we use more energy, discard more trash, and consume more goods. By no means am I suggesting that we have to give up all luxuries. It does mean we have to be more prudent in our choices. Socrates said many years ago, **"Be moderate in all things"**. Americans may have lost touch with that wisdom. The rising cost of fuel has forced many Americans to reduce their driving and find ways to reduce their energy costs. Fewer humongous SUV's and trucks are being sold and more fuel efficient hybrids are being purchased. As a result, the summer of 2008 fuel consumption in the US has decreased and the cost of a barrel of oil has dropped from its historical high. As the cost of fuel became more painful for millions of people, many Americans made individual decisions to reduce the amount of driving they did. This proves that when *individuals* like you and me take action we do have an impact on the overall problem.

As individuals we can do a lot to clean up the environment. In my seminars as well as talking to average people, I frequently hear the fear that "It may be too late." We don't know that. All we know is that if the planet and life as we know it has any chance of survival, we must start some time and some where. So why not start now?

The simple things we can change every day should be augmented by grass roots movements to prod our legislators, employers, and manufacturers at all levels to do more to protect the environment which really means we are protecting ourselves and our children. Otherwise, we can expect increasingly more severe weather patterns with a consequent change in the distribution of rain and drought. This will result in more difficulty growing sufficient food and higher levels of illness from cancer and respiratory diseases as pollution attacks our systems. The 2008 flooding in the Midwestern US destroyed millions of acres of crops. As a consequence, food will be more expensive, if available, to millions of people. Corn to produce ethanol used to offset high gasoline prices was one of the major crops destroyed so the price of gas could rise. It is clear that we are in a vicious cycle which we must break in order to effect positive change. It is also clear that we cannot expect politicians or big business to break this cycle. Working individually and collectively, *we* will have a more dramatic impact much quicker than any government agency. Besides, I like the idea of being part of the solution rather than the part of the problem, don't you? I have been part of the problem for too long and I can't fix the world alone. I know I can't change you, you have to decide to change on your own. Are you interested?

Figure 10. Melting glaciers.

Figure 11. Picture of low lying areas that may be
flooded as the ocean rises.

The alternative is to have the melting of the glaciers start to cover low lying areas and cities making many areas uninhabitable or radically changed. Major cities from Miami to Honolulu risk being lost cities just like the legendary Atlantis.

Polar bear habitat is being destroyed at such a rapid rate that they are now an endangered species as they are being forced to swim long distances to find food. On August 16, 2008, federal observers spotted nine polar bears swimming in the open sea. They were swimming north from the Alaska coast. The nearest land in the direction they were swimming was 400 miles! Even if they turned around and headed to Alaska some of them were still 65 miles to the closest land. (News. Yahoo.com, August 27, 2008). Do you think any of them made it to land? More likely they became so exhausted they drowned. Now, some politicians have decided that if a polar bear does happen to swim to their land that the polar bear should be shot as it is too expensive to re-locate them. How sad!

Figure 12. Exhausted polar bear.

To reduce the possibility of these things happening, we each have a responsibility to reduce global warming. This is the price we pay for eating, breathing, and living on Mother Earth. Let's work together to be sure that we and our children can THRIVE on Mother Earth!

So what can you do to help save the planet? The title of this book is "Resurrection of the Blue Planet" because the planet Earth can still be saved! *The key to reducing global warming is for each reader to find 3 ideas from this book that he can implement in his daily life.* If enough people do just a little, a significant impact can result. I've recently noticed that more and more people are using canvas bags at the grocery store and shunning the paper or plastic issue. In fact, some stores are even giving customers a discount if they bring in their own bags instead of using plastic bags. Gradually, more and more people are taking just one step at a time to change. We each need to break old habits (which will be hard), create new habits (which will be hard), and make those habits part of our daily rituals (which will be very hard!). You'll have to practice these three ideas faithfully for 21 days to make these habits into new energy efficient habits. Don't worry if you forget one time. Just keep trying every day and keep sight of your new habit until it becomes a part of you! Then add 3 more for another 21 days and keep gradually adding just three new ideas every 3 weeks. In less than one year you will have made a very significant difference in global warming. Individual super human efforts are not necessary but consistent small efforts will have significant effects. I am reminded of the question "How do you eat an elephant? Of course the answer is "one bite at a time!" Don't let the size of the elephant intimidate you as your job is to just take one bite at a time! Do what you can on that one day. You can't eat the entire elephant in one day and you can't cure global warming in just one day. The simple act of selecting 3 ideas will not mean that you will always remember to do what you have chosen to do. Initially, you will have to remind yourself over and over again to do your chosen idea. Sometimes, you'll forget to do it. Don't beat yourself up for not being perfect. No one is perfect. Just do it the next time and start over for your 3 weeks. The pages at the back of the book also ask you to have someone witness your personal commitment to do these three things. Your witness can also serve as your personal coach to encourage you, support you, and keep you on track. Ideally, you would also be the coach/witness for this person so you have a mutually supportive team. This is a buddy system used for fitness training in which one person drags the other one along when one person doesn't feel up to going to the gym. It is a symbiotic relationship so "nagging" is not allowed. ;)

This plan has the greatest benefit if you gather a group of friends, a church group, or a team of colleagues to each commit to making 3 changes every 21 days and in one year, we should start to see improvements. Make it fun! Make it a contest! Give out awards to the one who has done the most to preserve our lives!

3 Ideas for 3 Weeks +

3 Ideas for 3 Weeks +

3 Ideas for 3 Weeks X

Millions of People =

LESS GLOBAL WARMING!

Figure 13. The 3 idea approach.

Of course, the problem will not go away after just one year but it is a start. Once these habits have been engrained in your routine, you can continue to add new ways to reduce to your carbon footprint. There is nothing magic about doing three activities. ***The magic comes from doing something and making that something part of your daily routine.*** If you can only handle one activity at a time due to other concerns, it's ok to work on that one thing until you make it such a strong part of your life that you do it unconsciously. For most people, it requires a minimum of three weeks to form a habit so don't give up. You can do this.

Nobel laureate, Dr. Wangari Maathai tells a story about a tiny hummingbird carrying water back and forth in her beak to fight a forest fire. While the other birds and animals fled, this one lonely bird kept dropping her few drops of water on the fire and returning for more. Finally, one of the bigger birds asked the hummingbird why she was wasting her energy on a task that was clearly too big for her.

Figure 14. "I'm doing the best that I can!"

The hummingbird replied (without stopping), "I'm doing the best that I can". What would happen if all the birds and other animals worked together to put out the fire? Small actions by enough people equal big results! All that anyone can ever ask of any of us is to "***Do the best that you can!***" Doing nothing is not an option. You can't be faulted for doing your best and you never have to feel guilty for doing your best. When you fail to act or when you don't do the job to the best of your ability, everyone suffers. Your lack of action may be denying the ability of another person to breathe fresh air, to eat uncontaminated food, or drink clean water. Just a few small steps can lead to more small steps which lead to joining hands with like minded people.

I can testify that consistent effort by one person leads to new effort by others. When I lived in Hawaii, I walked past the Waikiki Marina almost every day and lamented how much trash was in the water. I felt bad for the fish that swam in this trash and embarrassed that the trash was the direct result of human neglect. So when the opportunity arose to volunteer to support my favorite local political candidate by cleaning up the harbor, I jumped at the chance. We cleaned the harbor one Saturday morning but I saw that it was filled with trash again the next day. I started going to the harbor each week and cleaning out trash. It was surprising how many people thanked me for doing this. Some would bring me cold drinks. Many people thought I worked for the county and were surprised that I didn't. I tried to do this every week end. After a while, individuals would stop to help including one who fortunately saved me from taking an accidental dip in the water as I slid off the wet concrete! One time while picking up trash along the water a high school senior had a camera and was doing an assignment on the trash in the water. He interviewed me and took a few photos. Not long after that, I saw a whole group of students picking up trash along that area of the island! So the concept works. One person can make a difference. One person can get other people involved by setting an example of doing what is right. I can't claim total responsibility for making this happen as other people also worked to clean the Marina.

Figure 15. Growing number of participants in fight against global warming.

Together we can solve this problem; however, we must have the cooperation of as many people in as many countries as possible to reduce greenhouse gases, replenish oxygen and fresh water, reduce harmful toxins, and to live in harmony with nature.

In spite of all the headlines and studies by scientists, there are still some people who deny the existence of global warming. To me the evidence is clear; however, I implore those individuals who question the evidence to ask themselves "What if the scientists are correct and we do nothing?

Will the problem become better or worse?" on the other hand "What if the scientists are correct and we are more conservative about our energy consumption? What will be harmed by being more moderate in our consumption of energy?"

By using the plan outlined in this book, becoming moderate in our consumption of energy is relatively easy. Setting specific tasks and monitoring your progress is very important as well as rewarding. With fuel costs soaring out of sight, the simple act of reducing your personal energy consumption will save you money! The entire family can participate and it can be fun too. There are a series of worksheets at the end of this book which ask you to choose three ideas, write them down, sign the worksheet, and get someone to witness the signing. By signing the worksheet, you are making a commitment to yourself to follow through on your selections. By having the signing witnessed by a friend, you are making a commitment to that other person that you will do it.

Calculate What You're Doing Now

Figure 16 Computer to calculate your carbon
footprint.

Ideally, one should measure how much energy they use in order
to verify that they are really reducing their energy consumption but it's
not necessary to get started. It stands to reason that if a person walks to
work or carpools one day a week, he has decreased his fuel consumption
over driving to work alone every day. Let's explore this one example. If a
person normally drives 20 miles round trip to work each day, he travels

100 miles per week for work. This translates into 400 miles per month. If his vehicle gets 20 mpg that means he uses 20 gallons at $ 4.25 per gallon or $ 85.00 for work travel. However, if he carpools just one day a week with three other people, he has cut his work travel because he will drive three days less per month or in dollar terms will save $ 12.75 per month in gas alone not counting reduced repairs, maintenance, tolls, etc. While this is not an overwhelming dollar figure, if you multiply it by the four people in the carpool, collectively your group has saved $ 51 a month or $ 612 per year. This translates to 960 gallons of gas not used by your group in a year and the reduction of hundreds of pounds of carbon emissions! If 100,000 groups did this, 96, 000,000 gallons of gas would not be used every year or a savings of $ 408,000,000. Do you think this would cause a bit of concern for OPEC? More importantly, there are fewer carbon emissions and the quality of our air has to improve.

We've mentioned the term carbon footprint several times. It is the amount of carbon emissions a given person or family adds to the atmosphere which lingers and contributes to global warming. It is usually expressed in tons as the average American family contributes a lot of carbon to the atmosphere. It is just like the footprint you leave in soft or muddy soil when you walk on it. Sometimes your footprint in the soil will disappear in a short period of time while in other cases it may last a long time. I recall when I was growing up on a farm that sometimes I would drive the tractor across a field when the clay soil was too soft and moist. The tractor would bog down in the clay and leave large ruts in the soil surface that would persist for years unless it was intentionally smoothed out. In some cases, the footprint you leave behind after you pass may be more like walking through soft concrete as it may take a very long time for the footprint to disappear. Our farm had very poor soil which had been over-farmed so that there was very little rich topsoil in comparison to the rich black topsoil filled with humus in the farm less than a mile away. As a small time farmer, I would envy the soil and the resultant corn crops my neighbor would grow on that rich black soil. Now, my heart bleeds as I visit that area and see the same soil that used to be rich and black is now hard and gray from growing the same corn crop year in and year out without replenishing the humus. That farmer still gets good yields of corn but he has to use too much synthetic fertilizer, pesticides, and herbicides to get a good crop. This is the tragedy that is being brought about on farms all across the USA. This is an example of this farmer's environmental footprint. The loss of humus rich soil and the addition of synthetic petroleum based fertilizer, herbicide, and pesticide. For most people, the carbon footprint is not clear because the carbon emissions they create are not easily seen. Your carbon footprint is the result of everything you do that increases greenhouse gases without doing anything to counterbalance or remove greenhouse gases.

There are a variety of internet sources that will help you calculate your carbon footprint. It's relatively easy using such sites as The Home Energy Saver or other sites (http://lbl.gov;www.cat.org.uk; www.climatecrisis.net) to estimate your carbon footprint. By calculating your carbon footprint, you can better identify your energy consumption patterns and identify potential places of change to reduce your carbon footprint.

This book is only a start and is a composite of ideas from many different sources as well as many of my own thoughts. I am confident that you can come up with more ideas that fit your personal lifestyle and implement them. Let's get started!

Conservation
It pays to be Green!

Figure 17. Conservation

As shown in the previous discussion, conservation is the quickest and least expensive way to reduce energy consumption and carbon emissions. Many conservation ideas don't require any costs and may yield immediate savings.

Low or No Cost (Immediate cost savings!)

1. Reduce your use of electricity
   a. Open the refrigerator, freezer, or oven as little as possible. When you do open the door, do as much as you can during that one opening. Unnecessarily opening the door or leaving open for extended periods of time wastes energy.
   b. Whatever device you have, if it doesn't require energy to function when you are not using it, turn it off.
   c. Use programmable or auto sensor safety lights to turn on outdoor light.
   d. Block off unused rooms by closing air vents, covering windows, and turning off A/C & heat vents.
   e. Use energy intensive devices like dryers and washing machines at night after peak energy hours. You'll usually pay less as energy companies often charge less during off-peak hours as they can reduce the number of plants they have to build and maintain.
   f. Hang clothes outside on a line rather than using a drier. They'll smell fresher without adding fragrances. If you live in a residential community, be sure to check the deed restrictions as some communities prohibit clothes lines.
   g. Operate appliances such as dishwashers, washing machines, and dryers fully loaded so you don't run as many loads. This conserves water, electricity, and detergent so you save all the way around.
   h. Replace all of your light bulbs with lower wattage and more energy efficient compact fluorescent lamps (CFL) which use up to 75% less energy, and last up to four times longer. You can save up to $ 25 a year or more per bulb by simply changing to a compact. As an added benefit, fluorescent bulbs sometimes seem brighter even at lower wattage. You can always use a different lamp shade to create a different ambiance in a particular room. Be careful not to drop or break a compact fluorescent bulb when changing light bulbs as some fluorescent bulbs contain hazardous materials. Some electric companies will provide these bulbs at low or no cost to their customers.

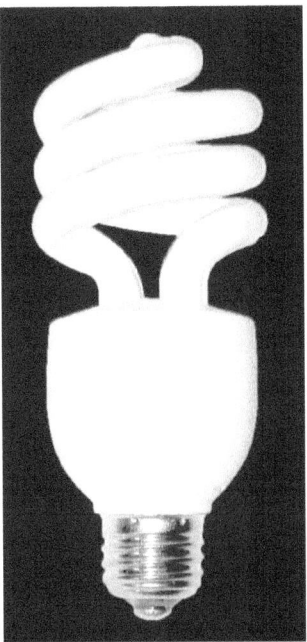

Figure 18 Compact fluorescent lamp (CFL)

i. Close draperies and shades facing the sun during the summer. Be sure draperies have a reflective color on the side facing the sun and are thick enough to reduce the passage of heat through the draperies.

j. Open draperies and shades facing the sun during the winter. This is especially helpful if the sun shines on a solid wall such as brick which will absorb the sun's warmth during the day and releases it back into the room at night.

k. Use a coffee carafe to keep coffee warm after brewing. It will taste better and use no electricity.

l. Use an under-the-sink hot water heater in areas where you just need a little water for occasional hand washing. These are available in most places where plumbing supplies are sold and are easy to install so most people who are handy with tools can do it.

m. Use motion detectors with outside safety lamps so the bulbs don't operate continuously.

n. Use large plants or small trees on balconies, lanais, and near patio doors to provide natural shade for the windows.

o. Check and repair magnetic insulation strips on your refrigerator and freezer for cold spots or signs of ice build up around the door for leaky insulation. The magnetic strip should be airtight all around the edges.

p. It is cheaper and more cost efficient to move air rather than cool it. Use a ceiling or other fan to move air which will allow you to feel cooler with a higher temperature set on the thermostat.

q. Hot air rises and cool air sinks so run the ceiling fan in the winter to push hot air down from the ceiling to where it will do more good. This is especially important in a room with a high ceiling or in southern areas where the heat ducts are in or near the ceiling.

r. Use fewer Christmas lights or use solar energy Christmas lights outdoors. Christmas and other holidays which have a long build up to a climax result in the greatest use of energy compared to other times of the year. Christmas shopping, Christmas decorations, wrapping papers, large meals, traveling to Grandma & Grandpa's house all contribute to higher energy expenditures. This fact does not mean you have to stop celebrating! Instead use your creativity to find ways to celebrate without requiring a whole new power station.

s. Open doors manually instead of using the button designed for the handicapped.

t. Turn off the TV and read a book or play with the kids.

u. Keep freezers at least 2/3 full as it takes as much energy to keep it cold when it's full as when it's empty.

v. Put a thermal blanket, newspapers, or a regular blanket over the food in your chest type freezer to keep the food cold if the electricity is going to be off for a few days. Avoid opening the freezer or fridge during these periods to keep things cold.

w. Place lighting where it will do the most good with the smallest use of electricity such as the corner of a room or opposite a mirror which will reflect the light back into the room.

x. Compare prices for electricity on the internet and check which companies offer electricity from renewable energy.

2. Changing your home temperature
   a. Adjusting your thermostat just 2 degrees lower in the winter and 2 degrees higher in the summer can save about 2000 # of carbon dioxide! Programmable thermostats allow this to happen without any conscious thought and can be installed by most people for less than $ 25.00.
   b. Adjust your thermostat at night to save energy.
   c. Hook up all electronic devices such as computers, printers, stereos and related equipment to a surge protector so you can turn them all off with just one button. (Be sure to shut off your computer with the proper procedure first).
   d. Keep your garage door closed to reduce heat or cold from entering the house and to serve as an air lock to mediate temperatures.

3. Conserve water
   a. Turn water off when brushing teeth, shaving, shampooing, etc.
   b. Be sure your automatic sprinkler is not watering your lawn during or immediately after rain showers, or during the day when the sun will evaporate most of the water. It's a good idea to water early in the morning or in the early evening.
   c. Use a low flow shower head.
   d. Taking shorter showers will reduce the amount of water used and the amount of energy used to heat the water.
   e. When taking a shower, get yourself wet, turn off the water while you soap up, and then turn the water on to rinse off the soap. This can substantially reduce the use of water and heating energy and save up to $ 500 a year in reduced energy costs.
   f. Use a low flow toilet to reduce the amount of water flushed away.
   g. Put a brick in the water holding tank of your toilet so it doesn't take as much water to fill it. Be sure that the amount of water in the tank will still flush the toilet completely. This is more appropriate for older toilets as most new toilets accommodate lower water usage.
   h. Don't flush the toilet each time you urinate.

i. Take showers instead of baths to reduce the amount of water and water heating energy used.

j. Collect rain water in a tank to be used to water your lawn and garden. A more sophisticated system could be built to allow rain water to be stored and used to flush your toilet or even filtered to be used for showers, or drinking water; however, drinking stored rainwater would require purifying so the average consumer would require professional help to develop a safe system.

k. Collect water run-off from your air conditioner and use it to water plants. In my condo, I have connected a hose to the outlet of the AC. This hose runs directly to the plants on my balcony so every time I use my AC I am watering my plants without additional energy or water. The whole system cost less than $ 10.00 and I don't have to worry about watering my plants!

4. Conserve Paper
    a. Just take one, rather than grabbing a handful of napkins or paper towels, just take the number you need. Remember a tree gave its life so you could wipe your mouth. ;)

    b. If you do get more napkins than you need at that moment, put them in your pocket or purse to be used later rather than throwing them away. I find it interesting that some fast food chains now keep the paper napkins behind the counter so the clerk can control the number of napkins dispensed. The result of this is that I never buy paper napkins because they always give me many more than I need. I don't ask for extra's they just give more than I need.

    c. Use both sides of the paper from your copier. Often, printer overruns or mistakes can be re-used for something else. For example, they can be cut into fourths and used as scrap paper for small notes instead of expensive "sticky notes".

    d. Put printer overruns and paper without sensitive information in a box and donate it to local preschools for the kids to draw on.

    e. Print out only what you need and save anything else electronically to conserve paper. Be sure to back up files on a disk or other external device in case your computer crashes.

f. Use a smaller font and more narrow margins on long documents to reduce the amount of paper.

g. Unfold boxes and store for future use. If you get too many large boxes, run an ad in a weekly paper or yahoogroups. com for someone who is moving. This is a form of the "pay it forward" concept popularized in a movie to encourage people to help each other.

h. Use a cloth or canvas bag for groceries so if they ask, "Paper or plastic?", you can say, "Canvas"!

i. Evaluate if you even need a bag for your purchase. If you don't need a bag, give it back to the clerk and say "No thanks, let's save the Earth."

5. Miscellaneous ideas

a. Try to reduce your weekly trash output by 10% each week for 5 weeks through conservation and recycling

b. Seek out conservation minded people with whom to share your "trash". Someone's trash may be someone else's treasure. In Honolulu, the website www.FreecycleHonolulu@ yahoogroups.com allows conservation minded people to post on-line items they would like to give away. Craigslists. com also has a recycling site. Anyone interested in the items can pick them up on a first come first basis and everything is free!

c. Pay all your bills and do all of your banking online. No more stamps, trips to the post office, and less paper consumed. While some banks do send out paper checks, many also do direct electronic transfers which consume less energy and no paper.

d. Use cash as little as possible to avoid having to go to the bank. Most places accept plastic and ATMs are everywhere. If you use your ATM card at a store for cash withdrawals, you don't pay a fee like you would if you went to another bank's ATM machine.

e. Decorate for festive occasions and holidays using re-usable decorations.

f. Dispose of hazardous waste properly. (Paint, batteries, old medications, acids, many household cleaners, bleach)

g. Recycle electronics. Almost all devices which have a circuit board may be recycled and electronic recycling centers are springing up all over. I used to give old equipment to my computer repair company who sold what they could as used and recycled the remainder.

6. Reduce your driving
   a. Work at home via the internet so you don't have to fight the traffic, burn up gas, and increase your blood pressure.
   b. Encourage the use of tele-conferencing rather than air line travel.
   c. Carpool
   d. Ask your company to set up four day work weeks so you work four 10 hour days and have 3 days off. This helps to conserve energy in many ways, may actually increase productivity, and makes life more pleasant to have 3 days without having to go to work.
   e. Ask your company to re-arrange when you come into work and when you leave to avoid bumper to bumper traffic jams which waste time and energy. For example, would it make a difference if your company started at 10 AM and worked to 6?
   f. Encourage local governmental agencies to develop carpool match databases with emergency backups in case your ride to work has to cancel the ride home.
   g. If your government won't do it, it could be a good project for Eagle Scouts or similar groups with members who are computer sophisticated (which usually means everyone over the age of 8 and less than 50) to set up a city or county wide carpool databases.
   h. If your company is large enough, encourage it to develop a carpool system. (You might meet a new friend from another department and not need that dating agency!) ;)
   i. Plan your errands so you can go everywhere you need to go but without crossing the same path multiple times. Try to set up your route so you only make right hand turns and you aren't idling in the left hand turn lane. This will be faster and safer too!

7. "Wash it, don't toss it!"
    a. Recycle plastic plates and tableware from fast food restaurants.
    b. Reuse plastic containers in every way possible
        i. Use as storage containers
        ii. Make them into toys for the kids. Be sure the recycled products are safe for kid's especially small kids who may swallow and choke on small items! Also, never re-use clear plastic.
        iii. Plant flowers in them
        iv. Know your plastics! Look on the bottom of containers and purchase products with #1 and #2 labels. These are easy to recycle and don't leave hazardous toxins in the environment. (AVOID # 7 or PC and #3 or V. These are made from bisphenol A and phthalates.)
        v. Avoid styrofoam as it does not de-compose! Do not heat food in a microwave in styrofoam containers to avoid contaminating your food from toxins emitted from the microwaving.

8. Reduce your use of disposables
    a. When purchasing a drink in a store, buy it where you can get a paper cup rather than buying a bottle, can, or styrofoam cup.
    b. Consume your drinks without a plastic straw. This "ancient technique" will require that you re-learn how to drink from a glass without the ice spilling everything on your lap. Believe me, it is possible. After a few attempts, even adults don't need "sippy cups" ;)
    c. Order your soft drinks without ice. You'll get more beverage and avoid the risk of e-coli infected ice. The drinks from the tap are usually cold anyway.
    d. Use a safe water bottle or canteen rather than a disposable bottle and refill using tap water filtered through a high quality in-home water filter such as Shaklee Best Water systems.
    e. Water bottles can be purchased online or in some sporting good stores which will filter tap water, are inexpensive,

and can be used over and over again. You won't need to buy bottled water. Read the labels carefully, however, to determine what the filters are removing and that the bottle is not made from a harmful chemical listed above. Also, be sure the filter is not adding harmful chemicals to your water.

f. Order most household products and many other items online in concentrated form and have them delivered. Evaluate if this is less expensive and energy efficient as it will be for some but may not be for others. There really isn't any point in buying products which are mostly water as you can add your own water without paying for shipping.

g. When you buy and store household products in concentrated forms, you don't have to buy as much or as often. Concentrated household products don't contribute as much to landfills and you don't have to drive to get them as often.

9. Reduce your use of hazardous and toxic chemicals by changing brands to non-toxic environmentally friendly household products in concentrated forms so you protect your family and reduce landfill trash. While many companies are now claiming to offer environmentally friendly products, the Shaklee Corporation has been a pioneer in such products for over 50 years and has been given numerous awards for their leadership efforts in protecting their consumers and the environment. Their Get Clean ™ product line will save you money, reduce landfill waste, protect your family, be as, or more efficient, than conventional highly advertised products, and protect the environment. They must be purchased through a Shaklee distributer or online at www. shaklee.net/wellnessstar. If you choose to use another product, be sure to read the label.

10. Teach others to conserve. Share this information with everyone and especially with children who are eager to learn and will benefit the most.

Plant

Figure 19. Gardening

Figure 20. Trees.

1. Plant trees which consume carbon dioxide and exhale oxygen.
    a. Memorial trees. Plant a tree in memory of a special occasion such as a birth, a death, wedding, anniversary, or birthday.
    b. Donate trees to your local school, park, or Habitat to Humanity
    c. Decorate your home, school, or office with trees which come in so many different varieties, shapes, colors, and blooms.

2. Plant an organic garden to reduce your dependence on store bought vegetables thereby reducing your trips to the grocery as well as the number of trucks needed to transport those vegetables to the grocery store. This is a great use for abandoned inner city lots as it builds community pride, improves nutrition to inner city kids, and reduces emissions.

3. Planting an organic garden also insures that your food has less likelihood of being contaminated with herbicides and pesticides. Unfortunately, most large city gardens will still have to contend with air borne environmental pollution which may leave residue on plants, fruits, and vegetables. Produce should always be cleaned before eating. Soak your fruits and vegetables for 5 minutes in a gallon of water with 4 drops of Shaklee's Basic

H 2 to remove such contaminants, then rinse. Basic H 2 is a surfactant rather than a soap so it will loosen and rid your fresh fruits and vegetables of contaminants.

4. Recycle water from your shower, washing machine, dishwasher, and rain gutters for your garden or lawn. Be careful to avoid using toxic chemicals like bleach or strong detergents which may contaminate your soil.

5. Plant shade trees to shield your home from the hot summer sun, to make your backyard more enjoyable, and to improve your property value.

6. If you live in the north, plant wind break trees to reduce the cold wind from chilling your house. Properly sited wind break trees may actually form a layer of insulation between the trees and your home to reduce heat loss.

7. Plant drought tolerant gardens. While all flowers need water regularly, many tolerate less water. You don't need to give up a beautiful yard.

8. Use ample mulch on your garden and around trees to reduce water loss and to build the soil.

9. Grass clippings work fine as mulch as long you haven't used an herbicide or insecticide on your lawn. You can use your own grass clippings and even pick up your neighbors thereby helping to reduce the landfills while improving your garden. Return your neighbors plastic bags for a refill after you've removed the grass clippings to further reduce landfill waste. When you use grass clippings for mulch, pile grass clippings and other organic wastes about 3 or 4 inches deep so there is aeration of the clippings for decomposition without odor. You can add more layers later. Be aware that grass clippings stored in a plastic bag get very hot through spontaneous combustion as they begin to decompose. Spontaneous combustion begins very quickly so never store them in your garage where they may start a fire. This is not a problem if you put them in the garden only if you pile them up in an enclosed area.

10. Use drip irrigation in your garden rather than a sprinkler to reduce evaporation and put the water where it will do the most good—near the roots.

11. If you do use sprinklers or a hose, water in the early morning or evening to reduce evaporation from the sun and to allow the

water to sink into the ground rather than allowing plants to get too moist which may result in disease or mold.

12. Plant a cactus garden if you live in the right part of the country so you don't have to water very frequently.

13. Instead of a lawn, plant ground covers that won't need as much water or mowing. Add drought tolerant flowers for added color.

14. Compost lawn, garden, and kitchen wastes. Inexpensive composting containers are available in many garden supply stores or supply houses such as www.Burpee.com. Some community colleges have composting workshops to teach people how to use their grass clippings, garden, and kitchen wastes. Some Starbucks offer used coffee grounds free to anyone who wants to re-cycle them in the garden.

15. If you don't have a garden, you can still compost kitchen wastes and junk mail in a red worm farm which will take up very little space. The compost can be given or sold to someone who does have a garden and the extra red worms can be sold to fisherman!

16. If you hand wash your dishes, don't run the water continuously. Wash some dishes in the basin and when they're clean, rinse them quickly at the same time.

17. If you hand wash your dishes, use the soap water from your wash basin or tub to clean the garbage disposal, water your plants, or flush your toilet.

Eating & Cooking

Figure 21. Save energy and money with cooking.

Cooking is an important task that may use a lot of energy. Careful planning of the menu and the cooking sequence can work wonders for reducing the amount of energy consumed. As a general rule, vegetarians use less energy overall and have a lower carbon footprint compared to meat eaters. Beef eaters will have the largest carbon footprint because it takes more time and energy to raise the cattle as well as to transport and refrigerate the meat. More importantly, however, cattle create a tremendous amount of methane gas which is one of the most powerful greenhouse gases. In addition, commercial meat farmers (which includes all types of meat and poultry) may use growth hormones and antibiotics to reduce feed costs while maximizing growth of the animals. As reported in the section on Toxins in the Environment, growth hormones have been implicated in a number of diseases while the antibiotics can help to create antibiotic resistant strains of bacteria. Some ideas are listed below to help reduce your carbon footprint while cooking.

1. Don't overcook your food. Steam your vegetables until tender but still crisp
2. Eat fruits and vegetables raw as they are better for you as you'll also get fiber, and the nutrients won't be cooked out.
3. If you do cook your vegetables, leave the skins on which will improve the nutrition of what you're eating, increase your daily intake of fiber, reduce landfill or disposal waste.
4. Eat brown rice (instead of white rice) which has been processed much less, has more flavor, more nutrients, and more fiber.
5. Use and eat whole grains rather than white processed flour.
6. Thaw foods before cooking to reduce cooking time. Thawing of meats and poultry should be done in the refrigerator to reduce the possibility of bacterial growth.
7. Use a microwave rather than an oven as it takes less energy to cook faster.
8. Lose weight. You'll need less food and will get improved gas mileage as your vehicle won't be carrying such a big load. You'll also be healthier and may reduce your risk of diseases such as diabetes, high cholesterol, and heart disease.
9. Lose weight so you'll have less fat as an insulation so your body will feel cooler.
10. Lose weight to reduce trips to the grocery store and the temptation it entails.

11. Lose weight as the less you eat, the better you'll feel. Weight reduction contributes to energy conservation, better health, fewer illnesses, fewer doctor bills, and fewer prescriptions. Furthermore, if you eat less, there is more for others with less energy required for production.
12. Use double boilers so you can cook two things at the same time.
13. If you're using the oven, check to see if you can cook several items in the oven at the same time.
14. Cook large portions on the weekends and freeze individual sized portions to take to work or school. This will allow you to save energy, save money, save stress, and eat less fast food (Saves money and improves your health!).
15. Barbecue less frequently to reduce carbon emissions. If you love the flavor of barbeque, why not cook more than you need at a given meal and freeze the extra for use at a later date. The smoky flavor will be retained in both vegetables and meats.
16. Wash and re-use plastic bags especially resealable bags.
17. Use plastic containers which have a longer useful life expectancy than using plastic bags (avoid BPA).
18. Buy in bulk from local markets or health food stores to help reduce packaging.
19. Encourage the re-use of frying oils which can be made into bio-fuels. More and more cities are making drop sites available for used cooking oil.
20. Re-cycle your fruit and vegetable waste products in a composter thereby reducing the amount of trash. The compost can be used in gardens or lawns. Composters are available from many large seed companies, do not attract rodents or insects and do not have an offensive odor.
21. Use paper plates and cups for outdoor or occasional use and toss into your composter instead of sending to the landfill.
22. Don't use a plastic cover or a straw for your fast food drinks.
23. If you must use a straw, encourage the use of paper straws or recycle plastic straws. One of my personal weekend routines is to remove trash from the marina near my home. It amazes me how many straws end up in the ocean and how difficult they are to remove because they slip through my net.

24. Re-use plastic water bottles. You can refill them at the tap or from a larger container of filtered water so you're not creating more trash. Some recycled plastic water bottles made from BPA plastic may leach harmful chemicals into the water you intend to drink especially if they have been scratched or if you use hot water to clean them. Read the bottom of the bottle for telltale signs of BPA. Do not re-use these bottles. Send them to be re-cycled and try not to buy them next time.

25. When cooking use the proper lid on pots and pans to retain heat, moisture, flavor, and nutrients. Foods will cook faster and taste better.

26. Turn off the heat when cooking or baking a few minutes (5-6) before the food is done. The heat stored in the pan, the food, and the stove top will usually provide enough energy to finish the job.

27. Foods that require a long cooking time may be cooked more efficiently in a pressure cooker to reduce cooking time, help retain flavor, and nutrients.

28. Reduce portion sizes so there are more portions per batch of cooking for freezing, to help lose weight, save money, and reduce waste.

29. Encourage restaurants to have "mini" plates with smaller portions which cost less and reduce waste. If the restaurant provides a large serving which exceeds your immediate interest or common sense, take the leftovers home in a doggie bag. This applies to those delicious desserts too. A bite a day of delicious cheesecake or chocolate éclair may not harm your diet but eating the whole thing at one sitting may be sabotaging yourself.

30. Reduce your consumption of red meat and poultry which consume a tremendous amount of energy to produce and emit methane gas. In addition, you may be reducing the frequency of your exposure to synthetic growth hormones and antibiotics used by commercial feeders to speed up the growth of the animal or bird.

31. Buy local, organic fruits and vegetables which do not use toxic herbicides, pesticides, and preservatives. Since they are grown locally, they require less transportation costs and emissions, help the local economy, and are a form of recycling. Wal-Mart made a

corporate decision to buy local produce to reduce transportation costs. This is a good step in the right direction.

32. If you take your lunch to work or on a picnic, old fashion wax paper does a good job, can be reused, and is less expensive. You can also reuse and recycle paper lunch bags.

Driving & Transportation

Figure 22. It's a long road.

Drive Green

The most important change you can make to improve your fuel efficiency and reduce your cost of driving is to change your driving habits. A fast start from a stop light may only get you to the next stop light faster and is a great way to throw your money at the pump. People buy different vehicles for different reasons but usually it's an emotional decision rather than a logical decision. Regardless of what we drive, the way we drive will determine how energy efficient the vehicle is. How many times have you heard someone say that their new car does not get the mileage estimated by the EPA? While the EPA system may be flawed, it does provide a consistent measure across vehicles for comparison purposes. The EPA system can't measure how well the car has been maintained or the driver's driving habits. For example, people who smash down on the accelerator at each stop light and jump from lane to lane will not get good gas mileage regardless of the vehicle they drive. Carpooling, driving with a gentle foot, and anticipating the road ahead will save money, fuel, frustration, and anxiety. In addition to the emotional toil of driving, there is considerable risk that the exhaust fumes inhaled while sitting in traffic can increase respiratory and other health problems. When you consider that some drivers annually waste a full week or more (40+ hours) stuck in traffic, it seems prudent that we make changes in our habits to make life better for everyone.

1. Park your car and walk to where you need to go can be a healthy pleasant experience. You'll save on your gas bills, may not need that gym membership you bought but never use, and will improve your health (maybe your waist line too!)
   a. Avoid drive through lanes especially if see a long line ahead of you. Quite often there is no line inside the restaurant or bank. Allowing your vehicle to idle for 5 or 10 minutes is a waste of energy which takes money out of your pocket and contributes to global warming. About 70% of all business conducted by fast food restaurants involves the drive through lane. Think about how many gallons of gas are wasted sitting in line and all the fumes you're inhaling just sitting there.
   b. Park away from the store where there are plenty of empty spaces so you don't drive around and around or sit idling. The extra walk will do you good, you'll be less frustrated, and probably get done shopping faster. You might even like the extra energy you gain as your health improves from this simple exercise program.
   c. Walk to the corner store or the mailbox or to take the kids to school
   d. Walk to work if feasible.

2. When sitting in your car in a parking lot, open the windows and turn the car off to avoid burning gas thereby reducing pollutants to the air.
3. Use a gentle touch when accelerating from a dead stop and gradually build up to speed.
4. Watch the traffic lights ahead to see if you can avoid having to stop again by timing your speed. Be sure to also watch any stopped traffic at the light to avoid a collision.
5. Encourage your elected officials to sequence the flow of stop lights so people don't have to speed up, stop, speed up, and stop, over and over again.
6. Learn the stop light timing sequence in areas you frequent and use that information to plan your driving pattern. It doesn't make sense to rush to a red light.
7. Take public transportation when possible and encourage the development of more efficient transportation (public & private).
8. Remove everything in your car that you don't need as it just adds extra weight which reduces your gas mileage.
9. Avoid excessive speeds. Drive near the speed limit so you can conserve fuel while not posing a safety hazard. For every 5 mph above 60, you are adding about 20 cents per mile to the cost of your trip. The increased fuel needed to operate at high speeds and wind resistance dramatically reduce your fuel efficiency.
10. Maintain your vehicle
    a. Change air, water, and gas filters as recommended by the manufacturer.
    b. Check tire pressure weekly to keep them at the proper levels. Underinflated tires can reduce your mileage by 1-2 miles per gallon and increase tire wear so you'll need to replace your tires more often. By checking and correcting your tire pressure, you'll save money on gas and tires will last longer so there will be more money in your pocket.
    c. Have your vehicle lubricated as recommended to reduce rolling resistance from bearings.
    d. Have catalytic converters checked and replaced if necessary.

e. Have your oil changed as recommended. Dirty oil may reduce your mileage by up to one mile per gallon.

f. Have annual emission tests and fix problems.

g. Have your wheels aligned to reduce tire wear and rolling resistance. If your tires are out of alignment, they don't roll as well so it may be similar to pushing the front tires down the road like sleds instead of rolling along. This may reduce your gas mileage by up to 2 miles a gallon.

h. Maintain your fuel, ignition, and emission systems to insure maximum fuel efficiency. One bad spark plug may decrease your mileage by up to 3 miles per gallon.

i. Be sure to tighten your gas cap after filling up to reduce gas evaporation. Missing or loose gas caps may result in almost 150,000, 000 gallons of evaporated gas per year! (Maui News, Sunday May 11, 2008).

j. Repair slipping automatic transmissions which can reduce mileage by up to one mile a gallon.

k. Replace cooling system thermostats which allow the engine to run at the wrong temperature. This may improve your gas mileage by several miles per gallon.

l. Use the proper grade of gasoline for your car. Be sure to check your instruction manual to see if a lower grade can be used without harm to your engine or performance. If the manual allows you to use a lower grade, you'll save money at the pump.

m. Use a good quality gasoline which has detergents to keep your engine cleaner for better operation and fuel efficiency. "Cheap" gas could be costing you a bundle in reduced mileage and increased repairs.

n. Keep your vehicle washed and waxed to reduce wind resistance. A clean smooth surface on your vehicle will allow air to flow over it more easily.

5. Plan your auto trips to minimize the number of miles you drive.

6. Plan your trips to maximize right turns so you don't have to wait for oncoming traffic to make a turn.

7. Don't set yourself up for a hard braking. Drive defensively and anticipate possible danger so you don't need to evade.
8. Carpool
   a. To work & shopping
   b. To drop off the kids
   c. To everywhere feasible

9. Ride a bike to work or for errands.

Figure 23
Biking can be fun, healthy, and help reduce global warming.

10. Consider driving a moped or motorcycle to work.
11. Lose weight. You'll need less food and will get better gas mileage as your vehicle won't be so heavy. Imagine if you have 5 people

in your car and they each weigh over 200 pounds. That means your car is carrying half a ton of weight.

12. Be patient with a slower vehicle ahead of you if your turn is coming up soon. The extra second or two you might save by passing them may cost you a lot of gas and could cause an accident so you'll be late anyway!

13. Rent environmentally friendly cars such as hybrids or those with better mileage. Enterprise Rent-a-Car is adding such hybrids so ask for them. In the near future, we should be seeing more electric cars which will be excellent commuting vehicles.

14. Have your vehicle checked if there is any visible smoke from your exhaust pipe as even diesel engines should not show billowing clouds of smoke or allow you to smell fumes

15. When booking air line travel, Travelocity's Go Zero will donate to the Conservation Fund to plant trees to counter airplane fuel consumption.

16. Expedia.com works with TerraPass to help fund greenhouse gas reductions.

17. Pick green hotels which are energy efficient such as Marriott, Fairmont, Kimpton, and Wyndham.

18. Check out GreatGreenTravel.com.

19. Cooperate with hotels and motels when they ask if you would allow the sheets to stay on your bed for your entire stay. After all, you don't change your sheets at home everyday, do you?

20. When you decide to replace your car, get one which has better mileage than the one you're trading in. More hybrid vehicles are now available but many non-hybrids have better mileage than older versions of the same model.

21. Remove roof top carriers, bike and ski racks that you aren't using as they just add to wind resistance. If you can't remove them, try to move the cross beams toward the back of your vehicle where they will be less prone to interfering with vehicle aerodynamics.

22. Remove bug and wind deflectors which also increase wind resistance thereby reducing fuel efficiency.

23. At low speeds, open your windows rather than use the air conditioning.

24. At high speeds, close your windows and use the air conditioning as the open windows increase the aerodynamic drag and reduce fuel efficiency.
25. Encourage your legislators to pass higher fuel efficiency standards.
26. Most large cities require that all vehicles pass an air quality test or smog test. When you see a vehicle that is emitting smoke from the tailpipe, take down their license number and report it to the police. Some cities have special numbers to call to report such violations while others may accept calls at the local station. Any vehicle (cars, trucks or buses) that is running properly should not be emitting visible smoke from the exhaust pipe.

Your Home, Building, Remodeling

Figure 24. Your home?

Your home is most likely the place where you use the greatest amount of energy and produce the greatest amount of waste. Many of the suggestions listed under **Conservation** impact your home energy use and the items below are more ideas to help you reduce your carbon footprint. Some of these ideas may cost less than $ 20.00 but others may run into the thousands of dollars. Keep in mind that there may be local, state, and federal tax rebate dollars available for larger projects.

For example, Hawaii Electric Company provides a $ 1000.00 rebate for installing solar energy. Other electric companies may have similar rebates or offer other help with replacing old appliances with Energy Star appliances. Before beginning large projects, it's a good idea to do some research on the internet for your state or to ask your tax preparer. The energy saving ideas for your home may pay benefits in many ways including greater comfort, improved aesthetics, reduced energy bills, increased property values, and the reduction of your carbon footprint.

Conservation

1) Insulate your home to a higher R value than the standard minimum recommended levels for your area. Be sure to include the ceilings and walls.

2) Insulate hot water heaters with an inexpensive hot water heater blanket (about $ 20.00) available at most building supply stores.

3) Insulate hot water pipes especially if they run a long way from the hot water heater. Your local building supply may have several options for this and most of them are inexpensive and easy for the do-it-yourselfer.

4) Replace single pane windows with double pane energy efficient windows. If it is too expensive to replace the windows, heavier drapes may also accomplish the same energy savings but you will lose the natural light when the drapes are closed.

5) Caulk around windows and doors to reduce air leaks. Large openings can be closed with expandable foam available at building supply stores: however, it may be more prudent (from an environmental toxin standpoint) to fix the hole and then caulk.

6) Be sure thresholds under the door are also not leaking air. Using a flash light from one side of the door may help. If you can see light, you have an air leak.

7) Check and repair air leaks around electrical outlets on outside walls and doors. Add weather stripping or fill large openings with expandable foam insulation.

8) Install solar energy to reduce or eliminate your electric needs. Some electric companies will give you a rebate for installing

solar panels. In addition, some state and federal programs may provide tax credits or rebates on energy conservation measures such as solar energy.

9) Add heat reflecting insulation in the ceiling. These panels are installed on the inside of the roof to help reflect heat energy.

10) Change air filters regularly so your furnace or AC doesn't have to work as hard. Also vacuum your air vents once a month especially during high use periods such as summer air conditioner use or winter heating.

11) Request air filters which can be washed rather than replaced. Most consumer home air filters are made of inexpensive plastic fibers that trap dust. It is perfectly conceivable that these filters could be made of natural materials such as cotton which could be washed and replaced.

12) When feasible, open the windows to allow air flow to help cool your home.

13) Add ceiling fans to move air. In the winter, ceiling fans will drive the hot air from the ceiling down to where you benefit from it so you won't have to turn the thermostat higher or run the heating system longer. In the summer, moving the air in a room with ceiling fans may make your home more comfortable with a higher thermostat on your air conditioner. It's cheaper to move air than to heat or cool it as ceiling fans may operate at 1/10 of the amount of electricity of an air conditioner.

14) Choose the outside color of your home and roof to maximum advantage. Lighter colors in warmer climates will reflect heat and darker colors in northern climates will absorb heat.

15) Paint inside walls to make the best use of lighting thereby reducing your need for electrical lamps.

16) Be sure your attic is properly vented and use an attic fan to remove excessive heat.

17) Use a solar powered attic fan to vent your attic.

18) Use solar lighting for outside lamps.

19) Fix all water leaks and dripping faucets. A single drop every second can waste up to 2000 gallons of water a year. This is especially important with a hot water faucet drip as both energy and water are wasted. Hot water faucet leaks are more common as the temperature of the water will reduce the life of the faucet

gasket. Changing the gasket on a water faucet is very easy and may cost less than $ 5 dollars so most people don't need a plumber to do this. If you have never done it before, just ask for help in your local full service hardware store.

2000 Gallons a Year!!!

Figure 25. Fix those drips.

20) Install room sensors which turn off lights when no one is there.
21) Support renewable sources of energy such as solar, wind, and sea waves when choosing an energy company. Until these industries get better established and more cost efficient, they may cost a little more per kilowatt hour. Consider these costs as a way to reduce your health care costs as air and water pollution are probably making you sick!
22) Install your air conditioner where it is shaded as much of the day as possible to reduce energy consumption by up to 10%. There are several ways to do this such as placing it on the side

of the house where it is shaded by the afternoon sun, building a canopy over it, or planting (or setting potted) trees or shrubs around it.

23) Turn fans and air conditioners off when they are not needed (e.g., when everyone is gone for the day or while away on vacation).

24) Be sure to close doors and windows if the air conditioner is running.

25) To help stabilize the overall home temperature, do not place the central thermostat near sources of heat or cold such as near a lamp, oven, an outside door or window or where it may be subjected to sudden temperature changes.

26) Turn your hot water heater temperature down to 120 degrees.

27) Have a plumber check an old hot water heater to see if it is filled with "junk" or calcium deposits which will rob efficiency. Hot water heaters over 10 years of age may be costing you money in wasted energy costs and may be worth replacing. This could save you up to $ 100 a year. When replacing a hot water heater, install a high efficiency hot water tank properly sized for your home. You need to heat enough water for showers, dishes, and laundry but not so much that you heat extra water which just gets cool again. Replacing a hot water heater may cost you about $ 350-550 so your payback time is relatively short.

28) If your current hot water heater does not provide enough hot water for you to finish your shower, it may be a good idea to check the cold water exchange pipe. This pipe delivers cold water to the bottom of the tank where it is heated before rising to the top. Old exchange pipes may be preventing the cold water from getting to the bottom of the tank. These exchange pipes can be changed quite easily and at less cost than replacing the entire tank.

29) If you live in an area with lots of sunshine, a solar powered hot water heater may save up to 80% of your energy costs for hot water.

30) Try to centrally locate your hot water heater as close as possible to areas where hot water will be used most often such as the laundry room, showers, dishwashers, etc. The farther the hot water must travel to be used, the greater the energy loss.

31) Open faced fireplaces are beautiful but they will drain your home of heat. If you love the ambiance of a fireplace, close the front with insulated glass doors to allow you to see the fire's glow. You can also add a blower to the fireplace to blow the hot air into the room so more heat is used for keeping the house warm rather than going up the chimney. Fireplaces will suck heat from your home even when they are not lit so it is an excellent idea to close them off with the glass doors.

32) Set thermostats at warmest comfortable setting in the summer and the coolest comfortable setting in the winter. It's fairly easy to wear light airy clothing in the summer to keep cooler and layer your clothing in the winter to keep warm. This can reduce energy costs dramatically and is an easy inexpensive solution.

33) Fully loaded dishwashers use less energy than hand washing dishes. Turning off the heat on the drying cycle to allow the dishes to air dry can save even more energy. In some locales, you might have to open the dishwasher door to allow the dishes to dry as the local humidity may be so high as to restrict air drying with the door closed.

34) Clean out your refrigerator regularly to discard those items which are just taking up space and using energy.

35) Once a month, pull your refrigerator out from the wall and vacuum the cooling elements at the back to remove dust which can reduce efficiency.

36) Use cold water for laundry loads if possible as this will reduce the use of energy for heating the water. You may not see any difference in cleaning power with cold water if you use an excellent concentrated laundry detergent such as Basic L from Shaklee ™.

37) Don't over-dry your clothes in the dryer. Be sure to select the proper dryer temperatures and use the proper amount of time for each load. Some fabrics don't require a long dry cycle while others (e.g., towels) require longer. When feasible, dry different fabrics in different loads. Newer dryers with "moisture sensing" controls can also help reduce drying times.

38) Clean dryer filters with each use to improve dryer efficiency and periodically vacuum the dryer vents to be sure they are not being clogged and restricting air flow.

39) Avoid opening the dryer periodically to add or remove items. This just allows hot air to escape and adds to drying time.
40) Adding wet clothes to partially dried clothes increases the drying time as the partially dried clothes just get wet again. Never add wet clothes to a moisture sensing dryer as then everything will appear to be wet and it may over dry some items.
41) Once you've started to dry your clothes, keep the dryer running with consecutive loads to use the retained heat rather than having to build up new heat with every load. If possible, start your clothes drying rotation with the lighter fabrics in the first loads followed by heavier fabrics later.
42) Office and apartment building may benefit from a roof top garden. In addition to using potentially "wasted space", the garden will help reduce energy consumption during the summer due to the extra insulation from the soil, reduce energy loss from the building and allow individuals to feel proud of their municipalities. If food is grown in the garden, then there is additional benefit from reducing energy costs from processing and transporting the food.

Building

43) Build a full or partial subterranean home which will be a constant temperature year round. Just a few feet below the ground surface, the soil stays at roughly the same temperature year around (60 degrees F, +/-10). Homes built below the ground (full subterranean) require very little heating or cooling energy. New sunlight tubes allow placement of sun anywhere in the home to keep the home feeling bright. Add solar collectors or a wind generator and you could be selling energy to the electrical grid rather than buying it. Partial subterranean homes may have part of the living quarters below ground and other parts above ground. Proper positioning of the partial subterranean home may allow for the morning and setting sun to determine the ambiance. The overall structure of the home could be covered with soil to improve insulation and provide fantastic landscaping opportunities. The biggest restrictions to such homes are a) overcoming our love for energy inefficient homes (i.e. large walls exposed to the elements)

which are subject to environmental changes; b) deed restrictions; c) the neighbors' initial reaction; d) water tables and preventing water seepage. Such homes would have obvious benefits in many parts of the country including areas prone to wildfires as well as those subjected to severe winters. Imagine an entire subdivision of subterranean homes which would have streets and landscaping, but no visible houses!

44) Build environmentally friendly. Find an architect who will design your home to fit the environment. Frank Lloyd Wright was a pioneer in such designs and some architects still use his ideas.

45) Depending on your where you live, try to build your home so that winter sun can be absorbed and stored in a heat retaining wall. As the sun goes down, the heat will naturally radiate back into the room.

46) Rather than cut down trees, plan your new home around the trees to maximize shade and improve the value of your lot. The old oak tree is very valuable during resale as well as to help you reduce your carbon footprint!

47) When building or remodeling use recycled materials. Many new construction materials made from recycled materials can be purchased. I remodeled my entire basement recreation room with recycled barn wood which I got for free. I even built a bar made from hand-hewn barn beams which were over 100 years old.

48) Build thicker outside walls to allow more insulation. This can be done when building your new home or could be part of a remodeling project. If remodeling, be sure to consult with an expert so you are not trapping moisture in the wall which will cause problems later.

49) Build into your new home air cleaners and water purifiers to help protect your family from air borne dust and toxins. This can be very important if you have someone with asthma or who's respiratory or immune systems are compromised.

50) Add or build into your home, energy efficient windows such as double paned tinted glass. A solar heat gain coefficient of .40 or less will give the best reduction of sun.

51) When replacing or building a floor, wood is an excellent material as it can last the life of the home, is a renewable-non petroleum based product, and gives a warm ambiance to your home.

52) Install only energy efficient appliances, furnaces, and air conditioners. Most appliances today have an Energy Star rating to allow you to make an informed decision about their energy use.

Figure 26

Energy star appliances can reduce your electricity use and save $ 100 or more per year per appliance. Some home economists recommend replacing your refrigerator and dishwasher if they are more than 10 years old to save energy.

53) Before replacing your refrigerator, you may want to see where energy is being lost. If there is obvious moisture or ice around the door, you know you have a problem. Otherwise, check the seals around the doors to be sure they are tight. To do this, take a piece of paper or 3 X 5 card and insert the card between the door and the seal. If you can easily pull the card out, the seal is not tight and energy is being wasted. Call the manufacturer to inquire about replacing the seals. Most of the seals are glued in place so it may be possible for you to remove the old seal and put new ones on. Next run your hand over the outside of the fridge and the freezer. If the outside is cold in some spots, then there is insufficient insulation in the door or wall. You may want to glue thick corkboard to the doors or walls in the cold areas. Double face tape or removable glue will allow you to take the corkboard off later. Corkboard will also allow you to put up notes and your kids' pictures on your fridge. You could also to put a throw blanket over the top and sides of the fridge. Throw blankets come in a wide variety of sizes, textures, and colors so you may be able to find something that matches your

décor. Another option to improve your refrigerator efficiencies is made by a company called Enigin (www.enigin.com) which is a British company that reports that 18 % of all electricity consumed worldwide is used for refrigeration. They market an energy saving device which they report has a short payback period.

54) Buy induction oven/ranges as they are more energy efficient than convection ovens. Induction ovens may use 90% of the energy to cook your food while convection ovens may use only about 20% so you must cook foods longer and use a lot more energy.

55) To improve energy efficiency and save your back, buy a self cleaning oven which may be better insulated so it won't need as much energy to use or clean.

56) Select appliances that meet your needs without being too large or too small. Refrigerators which are too large just waste energy while refrigerators which are too small become overcrowded and lose efficiency.

57) Unless you use a lot of ice, you can save energy by turning off your ice maker by raising the ice bar in the freezer. Ice cube trays still do a nice job for occasional cold drinks at a lot less cost.

58) Locate refrigerators and freezers away from ovens, dishwashers, or hot water heaters and place them so there is ample air circulation around the cooling elements for maximum efficiency.

59) Try to keep your refrigerator set for an internal temperature around 40 degrees F to keep food at the proper temperature. Freezers should be set at or near zero F.

60) Dispose of old appliances properly. If your old refrigerator was too costly to use in the kitchen, think carefully about using it in the garage and keeping it half empty. If you do dispose of an old appliance such as refrigerator, freezer, washer, or dryer, be sure to take the doors off before placing it out to be picked up. Kids have a habit of looking for places to hide and too many kids have been suffocated in old appliances.

61) Get expert advice before selecting the size of your furnace, heat pump, or air conditioner so you can buy a unit which will operate most efficiently and last the longest.

62) When purchasing a freezer, chest type freezers tend to use and lose less energy than upright freezers. It's also easier to put a blanket over them for additional insulation. Be sure not to cover the cooling elements with the blanket.

Apartment and Condo Dwellers.

Most of the conservation ideas proposed in this book can easily be used by people who live in apartments or condos; however, people living in these areas have special concerns as they don't have control of the operation of the building. There are some things they can do to reduce their carbon footprint.

63) Place tall bushy plants in pots on your balcony or lanai to shade your windows or your air conditioner.
64) If the building does not provide recycling bins for cans, bottles, newspapers, and plastics, the building supervisor may allow space for recycling if a trusted tenant or condo owner takes responsibility for coordinating the recycling.
65) While it should pose no problem to replace incandescent lamps with CFL in your home, replacing incandescent lamps with energy saving fluorescent bulbs will involve a major investment if the building is very large. The energy savings, however, will be significant and the payback will be very short. Some electric companies may subsidize the change-over for buildings that consume a lot of electricity. If the building supervisor has not changed the lamps in the common areas, suggesting it to him and showing how much he can save may help him to move forward on this.
66) A small group of interested tenants or owners may get together to do a survey and brainstorm on how to reduce energy consumption in their building. As long as the payback exceeds the expense, many supervisors will work with such an ad hoc committee to help reduce their expense.

Noise Pollution

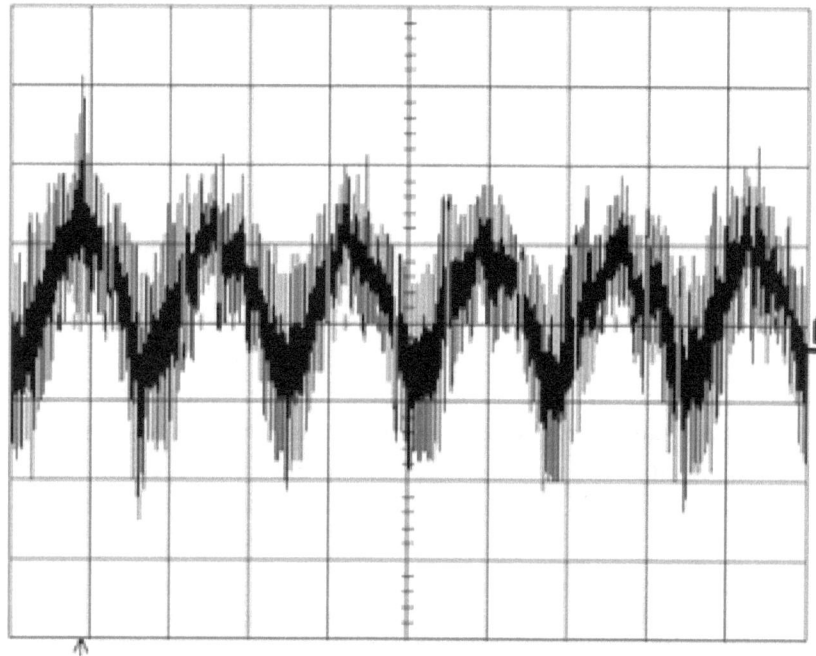

Figure 27. Noise bombards us all.

Noise is another major form of environmental pollution that affects people, animals, and our habitats. It damages hearing, increases mental illness, causes ringing in the ears, increases heart rates, changes hormonal production in the body, raises anxiety levels, reduces one's ability to do complex tasks, changes DNA, and has been implicated in the death of many varieties of marine animals. Noise is such a constant source of pollution in modern civilizations that few people really take note of it. Yet it is rare to find an area in or near a city where there is no noise. Without much conscious thought, we use noisy vacuum cleaners, dishwashers, lawn mowers, power tools, drive or ride to work in noisy vehicles, trains, or buses every day. While each individual exposure to noise may not be hazardous, no one really knows the cumulative effect of many exposures every day. Over 50 years ago there was a study done by Dr. Samuel Rosenburg who tested the hearing of people in New York City and compared their hearing to a group of African tribesmen called the Mabaan tribe. The Mabaan tribe lived in a remote area of Africa which did not have modern machines and all the consequent noises from industrial societies. Dr. Rosenburg found that as the New Yorkers aged, there was an increase in the prevalence of hearing loss. However, this was not found with the Mabaans whose hearing was the same for their young and their elderly. Dr. Rosenburg suspected that the high level of noise which bombarded the New Yorkers was at least one of the major causes of the increased hearing loss.

While the incidence of hearing loss affected approximately 8% of the US population 35 years ago, the incidence is now closer to 10% of the general population. Adults who would not be expected to have hearing loss until they became elderly are now showing hearing loss in unprecedented numbers and it has been estimated that up to 50% of Baby Boomers are showing signs of hearing loss. This is in sharp contrast to the historical figure of 33% of Seniors at age 65 having hearing loss. Baby Boomers were the first generation of Americans to grow up with loud music and noise exposure at home, in the workplace, and during recreation. Another recent study reported that 50% of teens were asking their peers to repeat conversation for them. Another report indicated that hearing loss is the most common disability affecting soldiers returning from the Iraq and Afghanistan wars. Without question, noise pollution is wreaking havoc on our ears.

Humans are not the only species to be affected by intense noise. It has been shown in a number of different studies that 1) some birds will not nest where the noise levels are too high; 2) whales have been known to beach themselves near areas where Navy sonar with intense noise has been conducted; 3) giant squids have shown damaged ears and internal organs from intense noise; 4) noise interferes with animal communication including mating calls and behavior; 5) noise has also been shown to disrupt the DNA sequence in mice. These are just a few of the effects of noise on animals.

In contrast with vision where one can close his eyes and block out what he sees, the ears can't be turned off and operate 24/7. Because of this, noise (or music) has been used as an aversive stimuli as well as in torture. For example, classical music has been used to discourage teens from loitering around convenience stores in Southern California and intense rock and roll music was used to drive former Panamanian strongman, Manual Noriega, from his compound.

How Loud is too loud?

Decibel	Sound	Allowed Exposure
150	Firecracker	NONE
120	Ambulance	< 3.75 MINUTES
110	Chain Saw Rock Concert	< 15 minutes
105	Personal stereo system	< 30 minutes
100	Wood shop snowmobile	< 60 minutes
95	Motorcycle	< 120 minutes
90	Power mower	< 4 hours
85	Heavy city traffic	8 hours
60	Normal conversation	No restrictions
40	Refrigerator humming	No restrictions
30	Whispered voice	No restrictions
20	Threshold of normal hearing	

Figure 28. Table of noise levels.

Music has also been successfully used to encourage shoppers to linger longer in stores so they spend more. So the subconscious implications of noise or music can be quite profound.

Anyone who has ever had a neighbor who played their music extremely loud at all hours of the night or had a dog who just won't stop barking is well aware of how this can be annoying and disrupt sleep. Audiologists are frequently asked to help select noise reducing ear plugs for individuals of significant others (both male and female) who snore because the snoring prevents them from sleeping! Sleep disruption can have a major impact on one's ability to function the next day and chronic sleep disruption can lead to depression and other mental illnesses. Patients who have chronic ringing in their ears (tinnitus) may suffer severe sleep deprivation, anxiety, depression, and other forms of mental illness because they cannot control the ringing and they can't get away from it.

Intense music is used regularly in aerobic and bicycle classes under the mistaken impression that the music stimulates the students to get a better workout. In actuality, loud music increases the heart rate by stimulating the release of the fight or flight hormones, adrenalin and cortisol. There was a form of hearing test used for newborns which presented various loud noise bursts to the newborn over a period of time. A special pad which detected the baby's heart rate was placed under the baby. Normal hearing was assumed if the baby's heart rate increased each time the noise was presented while babies who did not show an increase in their heart rate with the presentation of the noise were suspected of having hearing loss. The noise levels used in this test were not considered to be hazardous noise levels at the time and fortunately, this test is no longer used as better methods have been developed. The fact that heart rate increases with loud noise underscores the fallacy of using intense music in aerobic classes and may actually be counterproductive in building aerobic capacity. The flood of hormones triggered by intense noise results in a constriction of tiny blood vessels as the body shunts blood to organs necessary for fight or flight. These hormones can also produce other changes in blood chemistry. The body's response to loud noise is to release adrenaline into the bloodstream which can reduce weight loss and lessen cardio-vascular benefit. This may help to explain why many people who exercise frequently in high noise levels may not be losing weight.

Furthermore, the noise probably poses a hazard to hearing for both the instructors and students. The author has measured the music levels in a number of aerobic classrooms. In one case, the music level was 117 decibel (A) at a distance of about 20 feet from the speakers. In another study, the average noise levels in 30 different cycling classes was found to be around 111 decibel (A). Sometimes the peak noise levels in the classes were so loud they exceeded the level of the meter which was 148 decibel! The Occupational Safety and Health Act require mandatory hearing protection for noise levels at 111 dB when the time of exposure is longer than 15 minutes. Exposure to noise at 148 dB (A) is never a safe level. Such classes last an hour for students and longer for instructors who teach several classes in a day. Instructors often believe the students like the music loud when actually research has found most students found the noise annoying. Instead, they are getting a work out that potentially damages their hearing and reduces (rather than improves) their cardio-vascular benefits. While workplace noise is usually covered by State or Federal OSHA laws, enforcement of hearing protection laws may be lax and some industries such as farm workers may not be covered. In other areas such as night clubs, rock concerts, aerobic classes, and concert halls the laws may apply but may not have been examined in the context of noise pollution.

What can the average person do to reduce noise pollution?

1) The simplest solution is to turn it down. Anytime you have to raise your voice to be heard by another person, the noise is too loud! One of the disadvantages of regularly listening to loud music is that one develops a false sense of perception of what is a good music listening level. Music can be appreciated at a much lower volume level than it is often played. As previously discussed, intense music triggers hormonal changes and increases heart rate. These changes may become addictive so one may believe that "unless they *feel* the music", it's not loud enough. This is especially true with young people who are still learning boundaries and don't understand how they are harming their bodies. While much ado is correctly expressed about how loud young people play their music, concert music, dance music, and musical theater may be too loud as well.

2) Another easy solution is to use it less. It is well established that the body can tolerate louder levels of noise without damage if the duration of the exposure is shorter. Of course, this relationship between the duration of exposure and the level of the noise applies only to one specific exposure to that type of noise. No one knows for sure the effects of multiple exposures to many different noise levels.

3) Turning off devices which are not being used is an important overall conservation tool. If you don't need the device in a reasonable period of time, turn it off. This also applies to reducing noise pollution.

4) When you are in the market for a new appliance or device, consider the noise level as another factor in environmental protection. For example, many large home remodeling stores have displays which allow you to compare the loudness of bathroom fans measured in "sones". Sones are a measure of loudness comparing the loudness of a given sound to a standardized sound. The higher the sone number, the higher the loudness. Unfortunately, these measures are usually not available for all noise producing devices to allow the consumer to make an informed decision.

5) In situations where the noise levels can't be reduced, hearing protection is an important tool. For example, most target shooters today know that they should wear hearing protection when shooting. Hearing protection should also be worn in every situation where a person must speak up to be heard over the noise. This includes all forms of recreational noise, lawn and agricultural machinery, power washers, etc. Every time I measured the noise levels in my cycle class, I wore custom hearing protectors. As it would be unfeasible to wear noise reducing earmuffs in an aerobics class, custom hearing protectors will provide the same benefit, be less obtrusive and less cumbersome. A pair of custom hearing protectors will probably cost less than your designer shoes and last a lot longer. Most audiologists and hearing aid dispensers can make these hearing protectors for you; however, many pharmacies and hardware stores also sell inexpensive hearing protectors. While some of these may be used, cleaned and reused, many

will just add to landfill waste. Also, the inexpensive hearing protectors may not give the same noise reduction benefits as custom hearing protection.

For musicians and people who regularly attend loud music venues, there are special musician ear plugs that reduce the level of the noise while allowing you to hear all of the sounds you want but at a much safer level.

Figure 29. Musician earplugs
Custom musician earplugs are available in a wide variety of colors and noise reduction levels. (Picture courtesy of All American Mold Laboratories, Wichita, KS.

Figure 30

Richard Navarro, Ph.D.

Standard soft sponge earplugs are inexpensive and disposable. They offer good noise reduction if properly inserted into the ear. (Picture courtesy of All American Mold Laboratories, Wichita, KS).

Hearing protective ear muffs are another option and they come in a variety of colors and sizes. Some will fit more comfortably than others. Few provide good noise reduction if they are placed over eyeglasses or do not fit snugly.

Figure 31. Hearing protecting earmuffs

Hearing protectors are categorized by their "Noise Reduction Ratio" or NRR. A higher NRR number indicates greater hearing protection. It is rarely possible to exceed an NRR number of 30 for any single protector. In very intense noise levels, there may be some benefit of both ear plugs and ear muffs. All hearing protectors, however, must be worn and placed in the ear properly to provide benefit. It they are not in the ear canal all the way or if earmuff type protectors allow a leak around the edge, they may provide no benefit. Some people may use both earplugs and earmuffs at the same time, however, this does not give you double protection as very intense sound will vibrate the bones of your skull thereby directly stimulating the inner ear.

6) When using your cell phone, speak at a normal conversational level. Almost everyone has been annoyed by someone who is speaking very loud when using their cell phone. If the other person is having trouble hearing you, move the microphone closer to your mouth, move to a quieter area so there is not as much background noise to interfere with the conversation, or call the person back to see if a better connection will improve communication. Speaking at a normal conversational level may also help reduce your blood pressure as higher voice levels require greater exertion.

7) When purchasing a new vehicle, select one which also runs quieter and seems quieter when riding in the passenger compartment. One of the hallmarks of luxury vehicles is that they seem so quiet when operating at highway speeds. These vehicles usually have extra insulation around the passenger compartment to reduce noise levels which allows the passengers to arrive more relaxed. A recent news story reported that some of the electric hybrid vehicles may pose a hazard to blind people who depend on hearing to become aware of approaching vehicles. Hybrids at low speeds may be so quiet that blind people may not hear them.

8) Lubrication is another important way to reduce noise pollution. Anyone who has ever heard a squeaky door, fan, or other device should know that a little bit of lubrication may stop that squeak. A squirt of 3-in-1 oil will often work wonders.

8) Operating a device at a slower speed will frequently help to reduce the noise emitted or perceived as lower operational speeds may produce less noise as well as a lower frequency of noise. While lowering the frequency of the noise may make the noise less annoying, it does not necessarily reduce the potential damage to hearing or the environment as the Navy sonar testing has clearly demonstrated with it's ultra low frequency sonar.

9) The use of sound absorbing and non-reflective surfaces help to reduce the tendency for sound to bounce back and forth within an enclosed area with hard walls (reverberation). Such reverberation may increase the noise levels as well as disrupt communication. In your home the use of carpets, draperies, rugs, cloth furniture coverings, tapestries, sound absorbing or reducing panels will all contribute to a more pleasant noise level. In general, more massive and less smooth surfaces tend to reduce noise levels the best.

11) Another helpful and easy approach is to close the windows and doors in areas where there is a great amount of noise. This can help in urban areas where there is a great amount of traffic or construction noise. Closing windows and doors, however, may also increase energy consumption as air conditioner usage is increased so one must weigh the costs and the benefits.

12) Replacing single pane windows with double pane energy efficient windows will save heating and cooling costs as well as reduce outside noise levels.

13) Replacing hollow core outside doors with solid, insulated, or steel doors will also provide the same benefits as changing the windows. When replacing doors, it will also be important to be sure the molding or insulation around the door provides a tight seal around the door.

14) Many large rooms in office buildings are divided into cubicles to provide work space for individual workers. To some extent, these "cubes" also help to break up the sound from one area to another. The fact that they are open at the top, however, reduces their effectiveness as sound attenuators. Simple modifications in the construction of the panels can make substantial differences in their sound attenuation properties. For example, if the top part of the panels are raised upward another foot and extended toward the middle of the cube on all sides to form a type of tent which is open at the top, and increasing the absorption properties of the cube walls may make them more private and quieter.

15) A useful technique for reducing noise levels around highways is to properly place sound barriers along the highway so the sound is not allowed to reach residential areas. Concrete barriers and trees are often used to reduce sound levels. Trees are an excellent choice for such barriers as they help to reduce the noise, absorb

carbon emissions from the vehicles, and replenish oxygen. While concrete barriers may be used to reduce traffic noise and will do so faster than trees which may take a long time to reach maturity, concrete doesn't reduce the carbon emissions, supply oxygen, and is expensive to install. Concrete also will increase water run off from the side of the road while the soil around the trees will absorb much of the water. One solution is to plant both fast and slow growing trees along the highway. The fast growing trees will shoot up fairly quickly and help reduce the noise levels until the slow growing trees with wider branches and more leaves mature. Hardy evergreens will provide noise reduction year round even in very cold climates.

16) When a single noise source such as a motor produces the noise, it can be encapsulated in a container or noise attenuating walls to reduce the noise levels. Subway rail systems reduce urban environmental noise but may subject individual riders to hazardous levels of noise when used on a daily basis. In the home, it may be fairly simple to build a removable noise reducing panel over dishwashers.

17) Encourage the passage and enforcement of legislation to reduce noise levels. Some communities have noise ordinances which reduce the level of noise after a specified time at night. This could apply to loud parties, loud vehicles, or other loud noise producers. For example, in some cities airports are prohibited from flying planes after a certain time at night to reduce the noise levels for people living near the airport.

18) Educate people, especially young people, about the hazards of noise. This is one of the most important areas of noise reduction as it can bear immediate and future benefits. While the immediate benefit may be to allow the rest of the family or neighborhood to live in peace, the long term benefit will be to allow young people to preserve their hearing longer as well as serving as leaders for future generations in preventing the negative effects of noise pollution.

19) Noise levels should always be considered when building especially for potentially high noise projects such as intra-urban rail, subways, airports, and highways. Financial expediency at the time of development may create severe problems later as noise

levels become uncomfortable, annoying, or intolerable. The use of steel wheels on steel rails for intra-urban mass transit may be less expensive to install and maintain, however, the noise levels created can make the housing area around the rail tracks less desirable and drive down property values.

20) Encourage police to ticket or impound vehicles which have loud equipment (music, engines, or exhaust systems) or are driven in a manner as to present excessive noise levels. If there is no noise ordinance, most municipalities have disturbing the peace laws which may be used to reduce annoying noise levels.

21) Encourage enforcement of workplace noise laws in places not traditionally considered hazardous such as aerobic classes, night clubs, bars, concerts. If management does not voluntarily reduce noise levels when requested, an anonymous call to the local OSHA office may result in an inspection and possible fines which will usually get management's attention.

23) Encourage TV stations to play commercials no louder than the average levels of the typical show. Most stations will deny that this happens but most listeners know from personal listening experience that the commercials are usually louder than the programs.

24) Encourage movie theaters to play movie trailers at more comfortable levels.

21) Never play iPods, CD players, or stereos beyond ½ volume.

22) If someone else is wearing an iPod and you can hear the music from several feet away, encourage the user to turn the volume down for their own good.

23) Don't patronize places that have a history of high noise levels. Restaurants and night clubs are often such high noise environments. The combination of many people talking and the reverberation which results from hard walls, floors, and ceilings can result in very intense noise. These locations may present a double whammy to your ears if you drink alcohol. It has been shown that just 2 drinks of alcohol eliminates your ears' natural protective system so the loud sounds may to do more damage to your hearing.

24) Encourage restaurants and dining halls to use soft surfaces in the dining rooms to reduce sound reverberation and encourage sound

absorption. Acoustical engineers can help reduce the noise levels significantly and make the dining experience more enjoyable. In one large retirement community, the noise level created by several hundred people trying to carry on a conversation during lunch exceeded 148 decibel! Many restaurants have such loud, reverberant dining areas that patrons may have considerable difficulty enjoying the excellent food and conversation desired in a dining experience.

25) If you must raise your voice to be heard in your car, try to reduce the noise levels through insulation around the doors and windows.

26) Ask TV program directors and sponsors to keep or turn the noise or music levels so the speech or conversation level is 10 to 15 decibel above the background levels. This will enhance the dialogue as more people will be able to understand it.

27) Refuse to buy toys or video games which have excessive noise levels. Cap guns have become a thing of the past as they were shown to produce hearing loss but many video games are designed to give maximum sound levels. Noise levels can sometimes be controlled at the device so educating and monitoring children is very important.

28) Fireworks are another source of noise pollution and potential hearing damage. The use of hearing protection does not reduce the glamour of the fireworks display but can improve the safety. While fireworks are frequently used for a wide variety of celebrations, they pose many safety hazards and should never be used by children or other people who do not have the maturity and common sense to realize they are "playing with fire."

29) Require that mufflers be used on motorcycles, mini-bikes, snowmobiles and similar devices. The author has seen children as young as 5 years old with noise induced hearing loss produced by riding loud mini-bikes.

Toxins in the Environment

Figure 32. Poisons are everywhere.

One of the most insidious and hazardous problems with environmental pollution is the massive problem of toxic chemicals in the environment. While there is no question that many man made chemicals like plastic have done wonders for humanity, we are just beginning to see the tip of the iceberg as to the downside of so many chemicals. The average American is subjected to over 1500 hazardous chemicals every day which impregnate just about everything he touches including his food, water, air, furniture, bedding, and clothing. These environmental toxins can rob people of vitality, increase the risk of chronic diseases, and drive up health care costs. The tragedy of all these environmental toxins is that they often have long lasting effects in people and animals as they last a long time in the environment and their effects may not be revealed for years after the exposure. A recent University of Washington study examined top selling laundry products and air fresheners. All six of the products tested emitted at least one toxic or hazardous chemical yet it was not listed on the product label. Five of the six products tested emitted one or more carcinogenic pollutants which the EPA states has no safe exposure level.

As we approach the end of this book, I would like to share a personal experience. When my family lived in a small farming community in Iowa, we lived on a short street with only 5 houses at the edge of town. One day while I was in my garden a plane flew over very low to the ground spraying herbicide or pesticide on the corn field right behind my garden. As it past, I noticed a light mist on my skin and a bad taste in my mouth for a few minutes. Soon after that I learned that at least one person in each of the other 5 houses had developed or died of cancer. Our neighbor right across the street developed a softball size cancerous tumor in the period of just a few months and died. She was in her mid thirties when she left her husband and 3 young children. We moved shortly after that.

There are so many things we use in our daily lives such as plastics, household products, pesticides, pharmaceuticals, and industrial chemicals which have been implicated in many different illnesses, mutations or changes in developmental patterns and deaths that we must begin to examine the side effects and determine if the risks outweigh the benefits. For example, the early onset of puberty in young girls has been studied from a number of perspectives. There is a strong correlation that exposure to toxic chemicals beginning in the womb and continuing their entire lives from many different products results in early puberty. When young children are exposed to these products, they are exposed to a whole category of xenoestrogens which mimic the hormone estrogen and bind to the estrogen receptors in the cells. This may result in hormonal changes in the young girls to induce early puberty. For adults, exposure to these xenoestrogens have been implicated in the increase in breast cancer, prostrate and testicular cancer, and reproductive problems such as low sperm count. A University of New Hampshire study suggested a link between the rampant obesity in the US and flame retardant chemicals used in a wide variety of products ranging from sofas, mattresses, TVs, computers, toasters and others. These chemicals are inhaled, embedded in our clothes, and washed back into the ground water where they contaminate our food. Since they are fat soluble chemicals, they are stored in the fat cells in the body and have been implicated in causing stem cells to produce fat cells rather than muscle or neural cells. In the January 2008 issue of Lancet, environmental toxins have been correlated with increased insulin resistance which can lead to Type II diabetes. Another study from Mount Sinai School of Medicine examined the blood and urine of 9 subjects and found the presence of 53 chemicals known to cause cancer in humans or animals, 62 chemicals known to be toxic to the brain and neurological system, and 55 chemicals which cause birth defects. For over 30 years, the Environmental Protection Agency has been monitoring the fatty tissue removed during surgery and has found that 75% of all samples have 20 toxic compounds. This raises the question then, "Can we expect the government to protect us from toxins?"

It has been common knowledge for many years that pregnant mothers who smoke or drink alcohol increase the risk of birth defects in their children. Other research shows that babies exposed to toxic chemicals in the womb show an increased likelihood of immune deficiencies later in life. Furthermore, recent research now suggests that when the father is exposed to toxic chemicals, his sperm is damaged which can result in developmental problems in the father's children and grandchildren. The above information is just a sample of the studies which are leading to the overwhelming conclusion that environmental toxins are affecting everyone. Many years ago Rachel Carlson wrote her book "Silent Spring". Now we are at the brink of a much larger catastrophe due to environmental toxins affecting every creature on Earth. Global warming is a very real pressing problem and one in which we have all directly or indirectly contributed. Unless we also attack this problem of environmental toxicity, it won't matter much if the planet goes up in smoke as we won't be here to notice. Early in 2008, Senators Olympia Snowe and Ted Kennedy introduced legislation to prohibit the use of growth hormones and antibiotics for animals raised for food. This should be just the beginning as so much more must be done.

An Associated Press Investigative Report on April 19, 2009 reported that U. S. manufacturers including pharmaceutical companies dumped 271 million pounds of pharmaceuticals into American water ways. Many of these water ways serve as drinking water sources yet these chemicals are not monitored by local, state or federal authorities. Twenty-two detected compounds are listed by the EPA as industrial compounds or pharmaceuticals by the FDA. Manufacturers are not the only source of these compounds as consumers may be a major contributing source as some drugs are naturally excreted after passing through the body and others are flushed down the toilet for various reasons.

There are actions that you can do to help protect yourself and your family.

1) Buy only meat and dairy products which are certified hormone, pesticide, and antibiotic free.
2) Use glass utensils rather than plastic at every opportunity.
3) Read the labels of all household cleaning products in your home and get rid of everything which can harm you or your family. This will probably include everything you have under your kitchen sink. Proven safe, efficient, economical, and non-toxic cleaning products are available through the Shaklee Corporation which is has been the environmental leader for over 50 years. (*www.shaklee.com* or *www.shaklee.net/wellnessstar*)
4) Never use plastic, Styrofoam, or plastic wrap to heat your food in a microwave.
5) Buy and use natural fabrics such as cotton and wool for clothing and bedding.
6) Buy local organic products which do not have pesticides or herbicides.
7) Filter your tap water with a high quality water purifying system which removes toxic chemicals, bacteria, and fungus.
8) Contact your local waste management office about how to dispose of extra paint and other chemicals in the garage.
9) Become an advocate for improving the overall state of the Earth to help reduce global warming, reduce environmental toxicity, and improve the quality of life for plants, animal, and humans. There are a host of grass roots organizations which advocate for better environmental management. The Sierra Club is just one national organization that is doing this. There may be a chapter in your area.
10) Encourage international agencies to ban the use of toxic chemicals across the globe.

DO IT FOR THEM!

Figure 33 Kids from around the world are counting on you!

Go Get'em Tiger!

Figure 34 This tiger is ready to go.

*T*he single most important thing you can do to help save the *planet is to get involved!* This book lists just a few ideas to help you get started. Everyone should be able to find at least three ideas they can easily implement and keep at them for 3 weeks until they become a habit. If it is difficult for you to remember to do something, try tying a string on your finger or a rubber band around your wrist. If you use a brightly colored string, everyone will start to ask you why you have the string on your finger and you have a golden opportunity to be an Ambassador for global protection. So talk it up, Mr. or Ms. Ambassador! Working with your witness/coach under the buddy system is a great way to stay motivated.

Even young children can be taught to turn off the water and the lights when they leave the room. Teaching young children about how global warming may kill off the polar bears and penguins can have an important impact in their young lives.

Older children can be put in charge of recycling and may be able to start earning money by recycling cans and bottles. At 70 cents a pound, cash will accumulate over time. While they may not get rich, they will learn many valuable lessons such as how they are part of the fight against global warming, how to save money, and how to work for what they want!

Older children can also develop posters, science projects, and do environmental awareness presentations. An excellent resource for children of all ages is www.globalwarming101.com. This site has a complete syllabus which teachers can incorporate into their classes.

The following pages are worksheets to get you started and help you stay on track.

WE CAN DO THIS!

WE WILL DO THIS!

WE MUST DO THIS!

My Plan

My New Activities for the First Three Weeks.

I, _____, will begin today to implement the following three ideas to reduce my carbon footprint. When I have incorporated these three ideas into my daily habits, I will reward myself with _____.

1. _____

   _____

2. _____

   _____

3. _____

   _____

_____
Signed   Date

_____
Witness   Date

I AM A WINNER!   I CAN DO THIS!   I WILL DO THIS!

A journey of a 1000 miles begins with one step!

My New Activities for the Second Three Weeks.

I, _____, will begin today to implement the following three ideas to reduce my carbon footprint. When I have incorporated these three ideas into my daily habits, I will reward myself with _____.

4. _____

_____

5. _____

_____

6. _____

_____

_____
Signed   Date

_____
Witness   Date

I AM A WINNER!   *I CAN DO THIS!*   *I WILL DO THIS!*

Best way to eat an elephant is one bite at a time!

Figure 35. Elephant.

My New Activities for the Third Three Weeks.

I, _____, will begin today to implement the following three ideas to reduce my carbon footprint. When I have incorporated these three ideas into my daily habits, I will reward myself with _____.

7. _____

_____

8. _____

_____

9. _____

_____

_____
Signed   Date

_____
Witness   Date

I AM A WINNER!   I CAN DO THIS!   I WILL DO THIS!

There are no benches
on the road to success!

1966 Galien High School Class Motto,
Galien, Michigan

Courtesy of Harry Boyce and fellow alumni

Please don't stop here. Keep adding new ideas as they occur to you. Once you make a commitment you may find that you will think of new ideas that apply to your personal life. Also, incorporate these ideas into your workplace too. Every boss in the world is interested in saving money and energy costs are usually a big expense. If you are the boss, why not start a reward program for your employees to reduce energy costs.

A word about the author . . .

Figure 37. Dr. Richard Navarro

D r. Richard Navarro earned his Doctor of Philosophy from Vanderbilt University, a Master of Arts and Bachelor of Science, cum laude from Western Michigan University. He has been an elected official, an associate professor, researcher, writer, motivational speaker, audiologist, clown, and inventor. His love for the environment formed early in his life as his mother taught him to love plants.

While living in Chicago, he was mystified at the age of 8 when he witnessed how grass clippings buried under the soil resulted in huge potatoes grown in a small city garden plot. Later, his family moved to a farm where his love for the land and its stewardship was further nourished. He has been environmentally conscious and an organic gardener for over 50 years. Dr. Navarro's inventions include a device to produce electricity using renewable energy. Dr. Navarro is a member of Creative Minds Solutions, LLC, a research and design company. He is president of Blue Planet Energy, Inc. which is focused on alternative energy production and head of Wellness Star, an independent Shaklee distributorship.

It's in your hands

Figure 38

What are you going to do with it?

This page is for you to write down your own ideas that you can do to help reduce your carbon footprint. Be creative and have fun!

This page is for you to write down your own ideas that you can do to help reduce your carbon footprint. Be creative and have fun!